Earth Bible Commentary
Series Editor

Norman C. Habel

Acts

About Earth's Children:
An Ecological Listening to the Acts
of the Apostles

Michael Trainor

t&tclark
LONDON • NEW YORK • OXFORD • NEW DELHI • SYDNEY

T&T CLARK
Bloomsbury Publishing Plc
50 Bedford Square, London, WC1B 3DP, UK
1385 Broadway, New York, NY 10018, USA
29 Earlsfort Terrace, Dublin 2, Ireland

BLOOMSBURY, T&T CLARK and the T&T Clark logo are trademarks
of Bloomsbury Publishing Plc

First published in Great Britain 2020
This paperback edition published in 2021

Copyright © Michael Trainor, 2020

Michael Trainor has asserted his right under the Copyright, Designs and Patents Act, 1988, to be identified as Author of this work.

Cover image © borchee/istock

All rights reserved. No part of this publication may be reproduced or transmitted in any form or by any means, electronic or mechanical, including photocopying, recording, or any information storage or retrieval system, without prior permission in writing from the publishers.

Bloomsbury Publishing Plc does not have any control over, or responsibility for, any third-party websites referred to or in this book. All internet addresses given in this book were correct at the time of going to press. The author and publisher regret any inconvenience caused if addresses have changed or sites have ceased to exist, but can accept no responsibility for any such changes.

A catalogue record for this book is available from the British Library.

A catalog record for this book is available from the Library of Congress.

ISBN: HB: 978-0-5676-7294-0
PB: 978-0-5677-0377-4
ePDF: 978-0-5676-7295-7
ePUB: 978-0-5676-7296-4

Series: Earth Bible Commentary

Typeset by RefineCatch Limited, Bungay, Suffolk

To find out more about our authors and books visit www.bloomsbury.com and sign up for our newsletters.

In Memory of
Terry Trainor (1944–2018)
Denis Edwards (1943–2019)

Contents

Abbreviations	viii
List of Figures	ix
List of Illustrations	x
Introduction: Situating the Acts of the Apostles	1

Part One

1	Acts 1.1-5. An Ecological Orientation	19
2	Acts 1.6-11. The Ecological Mission	27
3	Acts 1.12–2.47. The Ecologically Renewed Household	35
4	Acts 3.1–6.7. The Fruitfulness of Earth's Children	43
5	Acts 6.8–8.1a. Earth's Presence in Stephen's Story of Israel	51

Part Two

6	Acts 8.1b–9.31. Water and Earth	65
7	Acts 9.32–11.18. Earth's Linen Sheet	73
8	Acts 11.19–14.28. Earth's Interconnectivity and the God of Creation	85
9	Acts 15.1–16.40. Earth Acts at Philippi	103
10	Acts 17.1–18.1. The God of Life and Breath	113
11	Acts 18.2–20.12. The Artisan, Artemis and the Lord's Supper	123
12	Acts 20.13–26.32. Earth's Child Identifies with Earth's Children	133

Part Three

13	Acts 27.1–28.31. The Final Voyage towards Rome and Earth's 'End'	143

Part Four

Conclusion. Luke's Ecological Resonances in Acts	163
Bibliography	173
Index of Authors	181
Index of References	185

Abbreviations

AEC	*About Earth's Child: An Ecological Listening to the Gospel of Luke*, by Michael Trainor (Sheffield: Sheffield Phoenix Press, 2012).
BA	*The Biblical Archaeologist*
BBR	*Bulletin for Biblical Research*
BTB	*Biblical Theology Bulletin*
CBQ	*Catholic Biblical Quarterly*
EDNT	*Exegetical Dictionary of the New Testament*, edited by Horst Balz and Gerhard Schneider (3 vols.; Grand Rapids, MI: Wm B. Eerdmans Publishing Co., 1990).
HTR	*Harvard Theological Review*
JBL	*Journal of Biblical Literature*
JSNT	*Journal for the Study of the New Testament*
JTS	*The Journal of Theological Studies*
LXX	The Septuagint
NIDNTT	*The New International Dictionary of New Testament Theology*, edited by Colin Brown (Exeter: Paternoster Press, 1975).
NovT	*Novum Testamentum*
NRSV	New Revised Standard Version
TDNT	*Theological Dictionary of the New Testament*, edited by Gerhard Kittel and Gerhard Friedrich, and translated by Geoffrey W. Bromiley (10 vols.; Grand Rapids, MI: Wm B. Eerdmans Publishing Co., 1964–76).
TS	*Theological Studies*
ZNW	*Zeitschrift für die neutestamentliche Wissenschaft*

Figures

1	Social stratification in Luke's Greco-Roman world	4
2	The intertextual process and textual conversation	7
3	Literary pattern of Lk. 2.14	25
4	Literary pattern of Lk. 19.38	25
5	Literary structure of Acts 4.32-5	46
6	The literary structure of the Stephen story in Acts 6.7–8.8	52
7	Saul's encounter with Heaven's Earth Child (Acts 9.1-19)	71
8	The thematic comparison of Peter's vision, as narrated in Acts 10.11-12 and Acts 11.5-6	78
9	The chiastic structure of Acts 10.14-15	80
10	The chiastic structure of the Philippian Event (Acts 16.12-40)	105
11	The theme of food that frames the ship saga (Acts 27.21-38)	147

Illustrations

1	Funerary stele depicting cart with rider and passengers, first to second century CE	68
2	Paul's first and second missionary journey	93
3	Roman ship with stowed amphorae, 75–60 BCE	94
4	Amphorae for transport and storage of food products	95
5	An inscribed funerary stele of a sailor in his small vessel	96
6	A section of the *Via Egnatia* in Philippi	103
7	Trajan's Column photographed around 1896	115
8	One of the panels on Trajan's column depicting the Dacians deforesting the Earth prior to battle	116
9	A tent made of goat hair in Wadi Rum, Jordan	124
10	Paul's third missionary journey	126
11	Artemis cult statue, second century CE	127
12	Artemis cult statue, first century CE	129
13	The remains of the ancient harbour of Caesarea Maritime	138
14	Paul's sea journeys including his final voyage to Rome	144
15	Paul's final journey to Rome, via Malta and Puteoli	153

INTRODUCTION

Situating the Acts of the Apostles

I live in the southern part of Australia. It is a beautiful part of the world, with fresh air, clean water, fascinating plants and unique fauna.

In the past month in writing these words, Adelaide, the capital city of the State of South Australia in which I live, experienced its hottest day ever recorded: in fact, it was the hottest city in the world at the time. Meanwhile, on that same day, in the north-eastern part of Australia, in the State of Queensland, 1-in-100-year floods covering 700 square kilometres devastated parts of the country, killing hundreds of thousands of cattle, and leaving farmers devastated.[1] These floods were so huge that they could even be tracked from space.[2] Further south from the Australian mainland, and on that same day, in the picturesque island of Tasmania, fires raged creating a 1,629 kilometre fire-front, consuming 190,000 hectares, equivalent to 3 per cent of Tasmania's total land mass.[3]

We who live in Australia do not need to look very far away from this island-continent to verify the conclusions of the 2018 report from the United Nations sponsored Intergovernmental Panel on Climate Change (IPCC):

> Human activities are estimated to have caused approximately 1.0°C of global warming above pre-industrial levels, with a likely range of 0.8°C to 1.2°C. Global warming is likely to reach 1.5°C between 2030 and 2052 if it continues to increase at the current rate... Climate models project robust differences in regional climate characteristics between present-day and global warming of 1.5°C, and between 1.5°C and 2°C. These differences include increases in: mean temperature in most land and ocean regions (*high confidence*), hot extremes in most inhabited regions

---

[1] ABC News, 'Queensland floods have likely killed hundreds of thousands of cattle, farmers facing "catastrophic" losses,' https://www.abc.net.au/news/2019-02-08/graziers-confronted-with-devastation-as-floods-kill-cattle/10793502 (accessed 20 November, 2018).
[2] *The Conversation*, 'Queensland floods are so huge the only way to track the floods is from space,' https://theconversation.com/queenslands-floods-are-so-huge-the-only-way-to-track-them-is-from-space-111083 (accessed 20 November 2018).
[3] Richard Flanagan, 'Tasmania is burning. The Climate Disaster Future has Arrived while they laugh at us,' *Guardian*, 5 February 2019; https://www.theguardian.com/environment/2019/feb/05/tasmania-is-burning-the-climate-disaster-future-has-arrived-while-those-in-power-laugh-at-us (accessed 20 March 2019).

(*high confidence*), heavy precipitation in several regions (*medium confidence*), and the probability of drought and precipitation deficits in some regions (*medium confidence*).⁴

Interpreting Biblical Texts Ecologically

Those of us involved with Christian theology and biblical interpretation cannot ignore this new context as we reflect upon our tradition and study biblical texts. Ecological issues and the care of the environment have become more pressing concerns, particularly in recent decades. Global warming is a reality. The *Earth Bible Project*, conceived in Adelaide, South Australia, seeks to offer one response to this.⁵ Scholars, engaging an 'ecological hermeneutic', interpret biblical texts from an ecological perspective.

This present book with its focus on the *Acts of the Apostles* is a logical sequel to an earlier ecological study of the gospel of Luke, *About Earth's Child: An Ecological Listening to the Gospel of Luke* (*AEC*).⁶ The approach I adopted in *AEC* will continue into the present, as I explore the ecological resonances throughout Luke's second volume.⁷

My intention in this introductory chapter is to present a brief background to what I consider Luke's context and agenda, and summarize the intertextual approach which I found helpful for listening to the gospel's ecological timbres. This same approach I bring to *Acts of the Apostles*. I also summarize other gospel insights that continue into Acts and are redolent with ecological emphasis. They sensitize the attentive listener to Luke's second volume and to the ecological possibilities that could emerge through an intertextual approach.

---

⁴ https://www.ipcc.ch/site/assets/uploads/sites/2/2018/07/sr15_headline_statements.pdf (accessed 3 February 2019). See also National Geographic's summary of the 2018 IPCC, 'Climate change impacts worse than expected, global report warns,' https://www.nationalgeographic.com/environment/2018/10/ipcc-report-climate-change-impacts-forests-emissions (accessed 20 March 2019).

⁵ For an overview and perspective on the Earth Bible Project, see Norman C. Habel, 'Introducing Ecological Hermeneutics,' *Lutheran Theological Journal* 46.2 (2012), pp. 97–105; Peter L. Trudinger and Norman C. Habel, *Exploring Ecological Hermeneutics* (Atlanta: Society of Biblical Literature, 2008); Norman C. Habel (ed.), *Readings from the Perspective of Earth* (Sheffield: Cleveland, OH: Sheffield Academic; Pilgrim, 2000).

⁶ Michael Trainor, *About Earth's Child: An Ecological Listening to the Gospel of Luke* (Sheffield: Sheffield Phoenix Press, 2012), hereafter referred to as *AEC*.

⁷ I do not presume here to write a full-blown commentary on Acts. There are so many fine commentaries which explore all the depth of Acts in a way that I would be challenged to do. A representative example of such commentaries upon which I rely include: Ernst Haenchen, *The Acts of the Apostles: A Commentary* (Philadelphia: The Westminster Press, 1965); F. F. Bruce, *The Book of Acts* (Grand Rapids, MI: Wm. B. Eerdmans Publishing Company, 1988); Hans Conzelmann, *Acts of the Apostles: A Commentary on the Acts of the Apostles* (Minneapolis: Fortress Press, 1987); Robert C. Tannehill, *The Narrative Unity of Luke-Acts: A Literary Interpretation, Volume 2: The Acts of the Apostles* (Minneapolis: Fortress Press, 1990); Luke Timothy Johnson, *The Acts of the Apostles* (Collegeville, MN: The Liturgical Press, 1992); Richard I. Pervo, *Acts: A Commentary* (Minneapolis: Fortress Press, 2009); Mikeal C. Parsons, *Acts* (Grand Rapids, MI: Baker Academic, 2008). These commentaries release me to offer a complementary engagement with Acts adopting an ecological hermeneutic, which I explain below.

There are also important ecological insights about Luke discussed in *AEC* that have implications for this study of Acts. I do not repeat these here. Rather, I refer the reader back to their discussion in *AEC* when appropriate. However, there are some matters raised in *AEC* which are important to revisit. These concern the proclamatory nature of Luke-Acts, the domestic context in which the proclamation was heard, and three other ecological features of the gospel that impact on Acts: Jesus' attitude to creation, Earth's presence (Greek, *gē*) throughout, and the richness of Luke's theology of God's 'reign', which I explain below as the *basileia-ecotopia*.[8] Finally I offer a schematic overview of Acts, concluding with some explanation about editing idiosyncrasies that I adopt.

Luke's Context

I presume, with most Lukan commentators, that the same author wrote both volumes and gave us, in Luke-Acts, the longest writing in the Second Testament.[9] While the provenance of Luke-Acts is speculative, its author sought to address the socio-cultural concerns of Jesus followers living in the later part of the first century CE, chronologically and geographically distant from their founding impulse in the story and practice of the Galilean Jesus.[10] Luke's writings came at a time of cultural immersion, as the Jesus movement, shaped by the spirit of the teaching and practice of Jesus of Nazareth, creatively engaged the urban context in which it lived. The movement chose this intentional cultural engagement rather than succumb to the temptation to build for itself a spiritual enclave untouched by the wider Greco-Roman world. As one Lukan commentator suggests:

> Limited understanding of Judaism and strong familiarity with the LXX [the Greek translation of the Hebrew Bible] suggest a gentile who had thoroughly immersed himself in Greek Scripture, perhaps a believer of long or even lifelong standing. Familiarity with the rhetorical technique and contact with such authors as Homer and Euripides suggest an education that progressed beyond the elementary level, but his

---

[8] In the Second Testament, 'Earth' (*gē*) occurs most frequently in Luke-Acts. One third of its occurrence is found in the Lukan corpus, with 25 out of 33 occurring in Acts. See Habel, *Readings*, p. 190, footnote 14.

[9] There exists almost complete scholarly consensus that the same authorial hand lies behind the Gospel According to Luke and The Acts of the Apostles. See, for example, I. Howard Marshall, *The Acts of the Apostles* (Sheffield: Sheffield Academic Press, 1992), p. 171; Joel B. Green, *The Theology of Luke* (Cambridge: Cambridge University Press, 1995), p. 47; C. K. Barrett, *Luke the Historian in Recent Study* (London: Epworth Press, 1961), p. 53; Charles Talbert, *Reading Acts: A Literary and Theological Commentary on the Acts of the Apostles* (New York: Crossroads, 1997), p. 3. But see the nuance of this authorial position summarized in Michael F. Bird, 'The Unity of Luke-Acts in Recent Discussion,' *JSNT* 29 (2007), pp. 425–48. 37,810 words comprise the Lukan corpus. See also, Mikeal C. Parsons and Martin M. Culy, *Acts: a Handbook on the Greek Text* (Waco, TX: Baylor University Press, 2003): 'Of the 5,436 unique vocabulary words in the New Testament, 2,038 are found in Acts (second again only to Luke with 2,055)', p. xiii.

[10] Scholarly dating of Acts is naturally conjectural! For an early dating (62–70 CE) see Karl Armstrong, 'A New Plea for an Early Date of Acts,' *Journal of Greco-Roman Christian Judaism* 13 (2017), pp. 79–110. Contra to Armstrong, I would posit a late first century dating for the writing of Acts, after the gospel was written and within the last decade of the first century CE. For a dating around the end of the first and the beginning of the second century CE, see Conzelmann, *Acts*, p. xxxiii. For an early second-century CE dating, see Richard I. Pervo, *Dating Acts: Between the Evangelists and the Apologists* (Santa Rosa, CA: Polebridge, 2006).

stylistic limitations indicate that he did not reach the advanced stages. Luke, as he is conveniently denominated, had at least occasional access to a wide range of Hellenistic Jewish literature. His cosmopolitan outlook suggests an urban background.[11]

Luke affirmed the culture of the day and recognized the need to reconfigure the gathering of Jesus members with an inclusive spirit, rather than conform to the socially stratified lines delineated by the cultural script of the day (Figure 1).[12] Wealthy disciples, the primary addressees of the gospel, were encouraged to release control over their possessions to ensure that poorer members were included fully in the community of disciples.[13] Material asceticism became a touchstone of discipleship. This ascetical

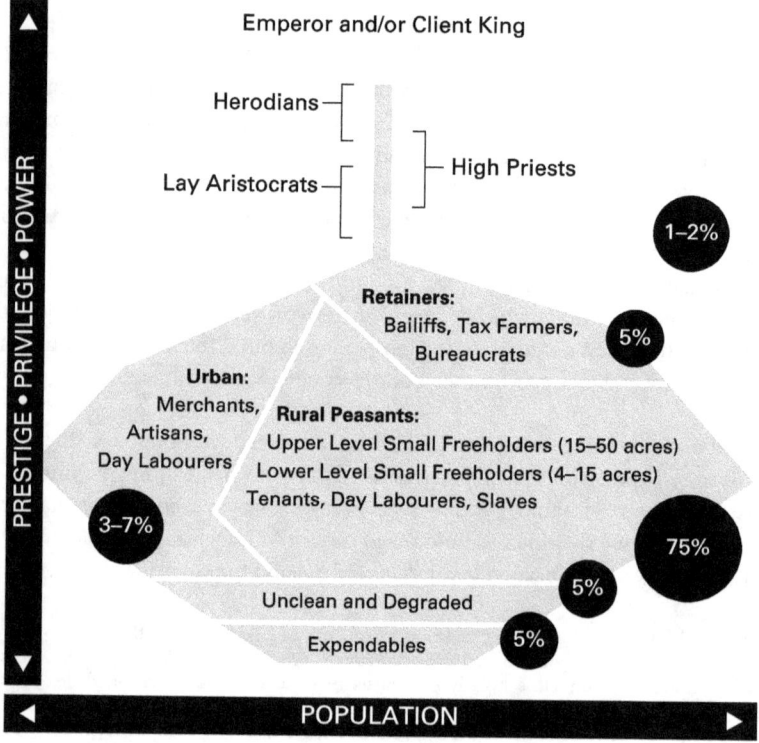

Figure 1 Social stratification in Luke's Greco-Roman world.[14]

---

[11] Pervo, *Acts*, p. 7.

[12] For a summary of this stratification, see *AEC*, pp. 27–9. For membership into the Jesus community by the more elite of Luke's audience defined by social stratification, see David Gill, 'Acts and the Urban Elites,' *The Book of Acts in its First Century Setting, Volume 2: Graeco-Roman Setting*, eds David W. J. Gill and Conrad Gempf (Grand Rapids, MI: William B. Eerdmans Publishing Company, 1994), pp.105–18.

[13] *AEC*, pp. 29–39 summarizes the gospel's background.

[14] Seminal to an earlier presentation on the ancient economy and peasantry is Moses Finley in Walter Scheidel and Sitta von Reden, *The Ancient Economy* (New York and London: Routledge, 2002). See the critique of Finley and other ancient economic theorists in Hans Derks, '"The Ancient Economy": The Problem and the Fraud', *The European Legacy* 7 (2002), p. 597.

spirit also had consequences for the way the gospel's audience treated Earth and regarded the natural world. A similar attitude surfaces in Luke's second volume, the *Acts of the Apostles*.

Luke confirms the missionary focus of Acts through the travels which Jesus' disciples undertake. The geography, which Luke presumes is panoramic, embraces the whole Mediterranean world from Cyrenaica (2.10) and Ethiopia (8.26-39) to Rome, from the most south-western and south-eastern corners of the Levant to the principal city of the world.[15] This world, the Earth, needs to be travelled around and over. It becomes the means or *vehicle* for the growth of the early Jesus movement as Luke writes about it. Geography is at the service of Luke's theology. Earth is central to this growth which happens, though not without its tensions and struggles.

Tensions surface in the story as the author reflects upon the necessity of incipient organizational structures for an audience that realizes that Jesus' final coming at the end of time (the *Parousia*) is no longer imminent. One example of this is the appointment of the Seven (6.1-6), established to address the overlooked needs of members of the Jerusalem Jesus community and to endorse the spread of the message of Jesus beyond Judaism. For Luke, the Jesus movement needs to consider sustaining structures for the 'long haul'.

Further, the relationship between Jewish and Gentile Jesus followers, especially in table companionship, surfaces in Luke's story. It is an issue that needs addressing. Though this is resolved in Acts 15.1-31, a perplexing reality for the evangelist was that most Jews did not embrace the Jesus movement. This bewilderment comes through Acts in different ways, especially in Paul's treatment from his co-religionists in their reaction to his preaching. More will be said about this when these scenes occur. But the fact that Luke usually has the itinerant Paul begin his urban preaching with a visit to the local synagogue indicates the writer's concern to address the Jewish world with the story of Jesus. The author seems puzzled about the overall lack of attraction to the Jesus movement by the Jewish majority. But the ongoing encounter which Luke's Paul had with his Jewish colleagues indicates the evangelist's affirmation of the ongoing salvific centrality of the Jewish people. The final scene in Acts, of Paul preaching to the Roman Jews, welcoming of him and yet divided about his preaching (28.23-31), mirrors the situation faced by Luke's late-first-century auditors.

Luke's Agenda

The overall agenda that drives Acts is the same as that of the gospel, to give assurance (Lk. 1.4) to Jesus followers in the later first century engaging the social realities of living in this Greco-Roman environment. Luke seeks to offer a social and cultural adaption of the story of God's reign revealed in, but not prescribed by, the ministry of the historical Jesus. Luke also wants to highlight key theological insights grounded in the

---

[15] Johnson, *Acts*, pp. 154–5. See also James Scott, 'Luke's Geographical Horizon,' *The Book of Acts in its First Century Setting, Volume 2: Graeco-Roman Setting*, eds David W. J. Gill and Conrad Gempf (Grand Rapids, MI: William B. Eerdmans Publishing Company, 1994), pp. 483–545.

ministerial praxis of the historical Jesus, explicated in the gospel and regarded as essential for later Jesus disciples.

For the contemporary interpreter of the gospel, a central theme of Luke concerns Jesus' relationship to the natural world and how we can identify a role which Earth plays in Luke's story. This links to the primary focus of this present volume. We can discern an ecological freshness in Acts that offers us encouragement in our concern about the planet and the use of Earth's goods for profit. As with Luke's gospel, it is possible to hear the evangelist's lengthy narrative in Acts with ecological ears attuned to the environmental resonances beneath the surface.

I repeat here what I wrote in *About Earth's Child*. Luke was not an ecologist nor did the author have concerns about environmental degradation. Certainly, there were perspectives on the use of the Earth, its soil and fruits, that were absorbed by Luke's audience. These perspectives also influenced gospel members in their perception of the Greco-Roman social ladder and its status stratification.[16] I have offered a hypothesis of the various attitudes to Earth in greater detail in *AEC* depending on whether gospel members were rural or urban peasants, artisans or the wealthy elite, and where they might be located on the Greco-Roman social grid (Figure 1).[17] Particular gospel characters – for example, the image of Mary in Lk. 1.26-38; the parable's crop owner and the gardener who protects a tree from being cut down in Lk. 12.16-21; the character of the luxuriant diner and the impoverished Lazarus in Jesus' parable in Lk. 16.19-31; or the many figures that appear in Jesus' teaching in the journey narrative in Lk. 9.51-19.27 – become symbolic representatives of Luke's audience subject to Greco-Roman stratification. Their characterization mirrors positive and negative attitudes to Earth. They reflect Earth-attitudes to which the contemporary disciple can be alert. This social, cultural and ecological background continues into Acts.

Approaching Acts: 'Intertextuality'

In *AEC* I offered an ecological hermeneutic in engaging the biblical text, and in particular the *Gospel According to Luke*. Without repeating in detail what I explicated there, I offer here some key points from *AEC* that lay the groundwork for the approach that I use in this present work.

Most biblical scholars come to the text concerned about how the text might speak to human beings and especially people of faith. This, of course, is important. The texts were written by people who encountered God revealed in the experience and stories of Israel (and expressed through the varieties of writings in the First Testament), and through the ministry of Jesus of Nazareth and those who gathered around him (expressed in the Second Testament collection). This overall agenda, of voicing the implications of this sacred encounter for believers, emphasizes salvation as an anthropocentric experience. It is what human beings experience and how they might respond to this experience.

---

[16] On Roman social stratification, see Gill, 'Acts and the Urban Elites,' pp. 106, 117; *AEC* pp. 26–30.
[17] *AEC*, pp. 30–8.

However, an unnoticed voice in the biblical world and texts is Earth. Retrieving the voice or presence of Earth in the stories and speeches in Acts is the foundational principle that undergirds my approach in exploring Acts. Sensitivity to Earth's presence, albeit camouflaged at times by the author's anthropocentric concerns, will allow a fresh way of listening to the ecological resonances within the text.

Allowing the environmental and Earth-related aspects to emerge through the story requires a careful and ascetical stance. This needs to acknowledge that Luke, a late first century Greco-Roman author, influenced by a particular time and culture, was not a twenty-first century westerner concerned about climate change. The author was one perhaps more in tune with the environment, agricultural practice, and geographical terrain and modes of travel than we are today. To repeat, Luke was not an ecologist.

Yet, the act of encountering an ancient text, like the *Acts of the Apostles*, is not a neutral, dispassionate one. The contemporary disciple brings their own perspectives to the ancient text. The encounter with the biblical text becomes a dialogical act, a sort of to-and-fro between the interpreter's world and that of the biblical author. Through this dialogue or textual conversation, meaning emerges (Figure 2). This, in theological language, is the 'Word' of God: a tensive, root metaphor that expresses the ambiguous human encounter with God's self-revelation as it is perceived in human terms. Personal insights and questions never remain constant throughout a person's life or history. The interpreter's world and their context are an important and necessary contributor to the interpreting process. This gathering of a person's life context and story creates a 'text' or a 'weaving': the Latin meaning of '*textum*' (from '*texere*', 'to weave') which is, literally, a 'woven piece of cloth' or 'web'.[18] This web of human interaction with the living, natural world of our lives creates the 'text' or con-*text* with which we encounter and seek to understand the relevant meaning of an ancient text.

Our personal contemporary 'text' engages the 'text' of the Bible in an intertextual dialogue and process (Figure 2). This means that there can never be a single, prescribed and eternally determinative meaning of a biblical text. In this dialogical, intertextual encounter fresh meanings can emerge in this engagement with the biblical text and the text of the *Acts of the Apostles*.[19] This also means that while the author, in our case, Luke, had a particular agenda in mind in crafting the narratives that form Luke-Acts, it is possible, indeed essential, that the contemporary world in which the interpreter lives offers a unique perspective with which one hears an ancient text in a way that is different, definitely foreign, to the original author.

The Biblical 'Text' ⟵⟶ Interpreter's 'Text': Contemporary Context

Figure 2 The intertextual process and textual conversation.

---

[18] D. P. Simpson, *Cassell's New Latin-English English-Latin Dictionary* (London: Cassell, 1962), p. 602.
[19] A brief word about the actual text of the *Acts of the Apostles* is necessary. There are two transmissions of Acts, the 'Alexandrian' (Egyptian) and 'Western' texts. Scholarly discussion continues as to the most reliable version. My approach is to follow the main opinion which gives preference to the Alexandrian recension. It is shorter than the Western text which shows signs of revision, stylistic additions and a tendency to explain expressions and situations. For a summary of the textual discussion, see Haenchen, *Acts*, pp. 50–60.

In the earlier explanation of this intertextual approach in *AEC*, the works of Julia Kristeva and other literary scholars were foundational for exploring the nature of a text and its dynamic.[20] They acknowledge in their studies the cultural and social influences that shape a text and the role which the interpreter plays in coming to understand a text's meaning for the present. This meaning might not be the original intention of its author. New insights will emerge in the dialogical-intertextual dynamic between original text and its interpreter.

In *AEC* I discussed the insights of the poet, T. S. Eliot, who reflected on the dynamic that occurs when a person encounters a poem.[21] He acknowledged the interplay between culture, interpreter and the literary text of the poem. In a 1919 essay, 'Tradition and the individual talent', Eliot asserted that a poem's quality was not the sole initiative of the poet. Its meaning resulted as a communal enterprise. Its language was formulated in the wake of how the poet's forebears understood their world and spoke about it. Eliot wrote, 'We shall often find that not only the best, but the most individual parts of [the poet's] work may be those in which the dead poets, [the] ancestors, assert their immortality most vigorously.'[22]

The 'individual parts' of the poem reflect the immortality of the poet's deceased forebears. Its meaning continues beyond the poet's life. It continues to offer new insights 'most vigorously' in subsequent generations. The implication of this for our approach to engaging Acts is obvious.

As contemporary interpreters, one of the 'texts' that concerns us must be our natural world and the environmental issues that face us. This 'text' becomes one of the dialogue partners I can bring to my reflection on the *Acts of the Apostles*. I listen for ecological and environmental resonances in the biblical text, not intended by its original author, but which I hear in the light of the present. I am convinced that the author of Acts would not have been aware of these, writing to a defined gathering of Jesus believers in a world with a certain environmental and geo-morphological stance from a Greco-Roman context of the late first century CE.[23] But the validity of the intertextual approach presented above legitimates me, the interpreter of the text, in bringing my ecological concerns to listen freshly to Luke's stories. This approach further decentres salvific anthropocentricism – the salvation of humanity – as the sole concern of the biblical author. It makes room for the natural world in which I live.

Insights from Luke's Gospel

There are other insights from *AEC* that are important in our encounter with Acts.

- The *Book of Acts* was meant for proclamation. Its congregational audience listened to it in an oral-aural engagement. Like the gospel, Acts was intended to be 'performed', that is, its proclamation by one appointed in Luke's Jesus gathering

---

[20] *AEC*, pp. 40–3.
[21] *AEC*, pp. 45–7.
[22] T. S. Eliot, *Selected Essays, 1917–1932* (New York: Harcourt, Brace & Company, 1932), p. 4.
[23] For the background and the audience, with their various attitudes to Earth, that I am envisaging that Luke addresses in Acts, see *AEC*, pp. 20–39.

allowed its verbal interpretative enactment to encounter those listening.[24] As in *AEC*, I prefer the expression 'auditors' rather than 'readers' to describe Acts' audience.[25] This act of listening makes room for potential ecological resonances to rise out of the text and to make connections through the ecological hermeneutic that I bring to this listening. 'Reading' the text, on the other hand, could force an interpretation onto the text that does not do justice to the author's intention. Listening allows the story of Acts to address the auditor who allows their ecological concerns to frame and resonate with the text.

- Those who listened to Acts gathered in 'households' of Jesus followers.[26] Churches do not exist at this early stage of the growth of the Jesus movement. Not until the Constantinian period and through Constantine's son, Constantine II (337–40), do formal ecclesial and basilica-style structures borrowed from civic architecture appear. The household context and liturgical setting for the oral-aural proclamation of Luke-Acts underscores the intimate relationships that characterized the Jesus gathering. Its socio-economic mix reflected and transformed the defined Greco-Roman social boundaries of Luke's world. Such gatherings would have certainly attracted suspicion from outsiders prepared to describe the Jesus members with pejorative expressions such as, for example, 'Christ-lackeys': a derogatory descriptor given to them at Antioch (11.26) because they were perceived as potential subverters of the Roman imperial system.[27]
- There are other features about Luke's writing that allow for contemporary ecological reflection and carry over into Acts:
  - In the gospel, Jesus' teaching to the disciples and their attitude to Earth, surface best in the journey narrative (Lk. 9.51–19.27). This section of the gospel encapsulates the central teaching for disciples influenced by attitudes of wealth accumulation, greed and the abuse of Earth's goods. Luke's Christology and the evangelist's presentation of discipleship, especially regarding ecological asceticism and environmental care, continue as an undercurrent in Acts.
  - Earth is present throughout the gospel. For example, in Luke's opening chapters, the evangelist presents Jesus at birth as God's agent surrounded by Earth's gifts, wrapped in bands of cloth and laid in a stone manger (Lk. 2.7, 12, 16). He is *Earth's Child*. In the gospel's final chapters, strategically framed by Mount Olivet

---

[24] *AEC*, pp. 48–9. Numerous studies have started to appear on biblical performance criticism. On this, see the contribution of David Rhoades, 'Performance Criticism: An Emerging Methodology in Biblical Studies,' https://www.sbl-site.org/assets/pdfs/Rhoads_Performance.pdf (accessed 12 February 2019). See also William Shiell, *Reading Acts: The Lector and the Early Christian Audience* (Leiden: Brill, 2004). Sheill argues that the text was meant for oral performance by a 'lector' to a generally illiterate Jesus gathering.

[25] *AEC*, pp. 43–8.

[26] Bradley Blue, 'Acts and the House Church,' *The Book of Acts in its First Century Setting, Volume 2: Graeco-Roman Setting*, eds David W. J. Gill and Conrad Gempf (Grand Rapids, MI: William B. Eerdmans Publishing Company, 1994), pp. 119–222.

[27] On the name, see Justin Taylor, 'Why were the disciples first called "Christians" at Antioch,' *Revue Biblique* 101 (1994), pp. 75–94. Also, David G. Horrell, 'The Label Χριστιανός: 1 Peter 4.16 and the Formation of Christian Identity,' *JBL* 126 (2007), pp. 361–81.

(Lk. 19.29; 21.38), concerned with Jesus' passion, death and resurrection, Luke confirms Jesus' Earth-connectedness. He shares a final meal with his disciples using Earth's gifts of wine and bread (Lk. 22.15-30). Later he comes to his knees, falls to Earth as he prays for strength and insight into what is about to befall him (Lk. 22.39-53). In his trial before Herod he is clothed with luminescent regalia (Lk. 23.6-12) which accompanies him to the cross of Earth's wood. He is affixed to this wood and it assists him in death (Lk. 23.26-49). Finally, as in the beginning of the gospel, he is wrapped in an explicitly identified linen cloth, a gift of Earth regarded as eternal and heavenly, and laid in a tomb carved out of Earth's womb (Lk. 23.50-53). The gospel is indeed a story of 'Earth's Child'. He attends to the Earth and its creatures; he is accompanied by Earth's gifts. And Earth does not disappear in Acts. As in the gospel, it is present in different ways.

Ecological Images in Acts

In the *Acts of the Apostles* Luke's focus shifts to those who are empowered by the Spirit of God to become the Spirit's agents as they move towards the 'the end of the Earth' (1.8c), preaching the gospel of the risen Jesus. This preaching is not only directed to humanity, it affects all creation.[28] The earliest members of the Jesus movement are Jerusalemites infused with their commitment to Earth's Child. They become the first Spirit-empowered agents of God's 'reign'. As seen in the gospel, Luke uses the term 'reign' (*basileia*) to express the direction of Jesus' ministry. His words and deeds reveal God's *basileia*, the presence and action of God that brings creatures into divine communion. The focus of this *basileia* is not purely anthropocentric. It includes all creatures, human and non-human, all matter, organic and non-organic, in fact, all Earth and everything that composes it. From this viewpoint, the *basileia* is also an *ecotopia*, an expression of God's ecological intent to embrace all creation.[29]

The *Basileia-ecotopia*

I continue to translate the language of 'kingdom' as *basileia-ecotopia* in the pages ahead, as Jesus' Spirit-filled disciples witness in Jerusalem, Judea, Samaria and 'the end of the Earth' (1.8c). The expression is central in Paul's teaching to his Roman audience in the final chapter of Acts which concludes Luke's story (28.23-31). Those who witness the presence of the *basileia-ecotopia* – the Jerusalem members of the Jesus movement, its leaders, especially Peter, the Gentiles whom Peter finally confirms as members, Paul and his various companions who accompany him in his travels around the Mediterranean in a particular part of the ancient world on Earth's surface – proclaim the message of Jesus, the Risen One and Earth's Child. Because of their communion

---

[28] The 'end of the Earth' is usually understood in terms of geography and the expansion of the gospel into the Greco-Roman world. See Daniel R. Schwartz, 'The End of the Γη (Acts 1:8): Beginning or End of the Christian Vision?,' *JBL* 105 (1986), pp. 669–76.
[29] On this expression *basileia-ecotopia*, see *AEC*, pp. 11 (footnote 13), 130.

with Jesus, Earth's Child, and their commitment to Earth as Jesus teaches them in the gospel, they, too, are children of Earth. As 'Earth's children', Jesus commissions them to travel across Earth's surface to Earth's 'end'. In Chapter 2, I explore a way of understanding the phrase 'end of the Earth'. I suggest that it is not only a cartographic descriptor; it is a multivalent ecological metaphor that influences the way we hear Jesus' injunction to his disciples to 'be my witnesses in Jerusalem and in the whole of Judea and Samaria and to the end of the Earth (*gē*)' (Acts 1.8).

The perspective of Luke's *Acts of the Apostles* as the story of Earth's children has inspired the title – *About Earth's Children: An Ecological Listening to the Acts of the Apostles* – and the following outline which falls into four parts:

Part One (1.1–8.1a) orients the auditor ecologically. It introduces us to the Jerusalem community of Jesus believers and acquaints the auditor with some of the ecological themes that will emerge throughout Acts. A central theme is the communion between Heaven and Earth. This conviction will appear often in Acts in different ways. Earth (*gē*) becomes an actor in many of the stories that comprise this section of Acts.

Part Two (8.1b–26.32) moves the auditor beyond Jerusalem and Judea into the wider Greco-Roman world of Asia Minor and Europe. Events around Jerusalem (Philip's baptism of an Ethiopian in 8.26-40) and Peter's healing activity in Lydda (9.32-35) and Jaffa (9.36-43) indicate that the Jesus movement will eventually move beyond Judaism and incorporate Gentiles into its membership. Conversion and openness to the 'unclean' Gentiles come to Peter through the power which Earth's gifts play in the form of a heavenly sheet (10.10-16), a centrepiece in the development of the Acts narrative. Peter's baptism of the Roman centurion, Cornelius, and his household (10.44-48) cements the incorporation of the Gentiles into the Jesus household. These events theologically endorse Paul's three missionary journeys as he advocates for the God of Creation and the influence which the natural world has on people's communion with his God revealed through Jesus. Luke's presentation of Paul and his various travels to the foremost cultural centres of Asia Minor, Macedonia and Greece occupy this major portion of Acts. Paul lives ascetically with an appreciation of the Earth and its seas upon which he travels. We might hear Paul inviting us to ecological conversion.

Part Three (27.1–28.31), composed of only one chapter, completes Luke's story of Earth's children. Paul arrives in Rome after a dramatic and event-filled sea journey. His preaching in his Roman domestic setting, itself an ecological space, centres on the *basileia-ecotopia* and Jesus' teaching. If he has arrived symbolically at 'the end of the Earth', the note upon which Luke ends this two-volume work of *Earth's Child* and *Earth's Children*, with Paul teaching 'openly and unhindered' (28.31b), suggests that this is not the end of the story. Nor the end of Earth. There is ecological hope.

Part Four is a final brief chapter. It summarizes the ecological links that emerge from an attentive 'listening' to the Acts of the Apostles. As with *AEC*, I conclude with summary theses that affirm the importance of ecological conversion for contemporary disciples informed by the biblical tradition and the *Acts of the Apostles*.

An Outline of the Acts of the Apostles

Part One

In Jerusalem: An Ecological Orientation (Acts 1.1–8.1a)

1.1-5	Prologue to Acts and recapitulation of the Gospel: Communion between Earth and Heaven.
1.6-11	The ecological mission: 'in Jerusalem, and in all Judea and Samaria and to the end of the Earth (*gē*)'. Jesus ascends: Communion of Earth's Child with Heaven.
1.12–2.47	The ecologically renewed household for mission: Luke's ecologically rich summary of the life of the nascent Jerusalem Jesus movement.
3.1–6.7	Fruitfulness of the Jerusalem Jesus household, despite frustration, opposition and greed.
6.8–8.1a	The Stephen episode: the agency of Earth (*gē*) in the story of Israel.

Part Two

Beyond Jerusalem into Europe: Ecological Conversion (Acts 8.1b–26.32)

8.1b–9.31	Water and Earth: Philip, the Ethiopian court official, and Saul/Paul.
9.32–11.18	A linen sheet and its creatures: the unclean declared clean; Gentiles become members of the Jesus movement.
11.19–14.28	The God of Creation: Paul and Barnabas. The first missionary journey into Asia Minor.
15.1–16.40	Earth acts: Heaven's communion with Earth redefines purity. Paul's second missionary journey begins: Asia Minor into Macedonia.
17.1–18.1	The God of life, breath and Creation: Paul's second missionary journey continues. Paul's Athenian speech.
18.2–20.12	The artisan Paul, Ephesus and Artemis, Troas and the Lord's Supper: a third missionary journey.
20.13–26.32	The ascetic Paul. Arrival back to Jerusalem. Trial in Caesarea. Third missionary Journey complete.

Part 3

To Rome and the 'End of the Earth': Ecological Hope (Acts 27.1–28.31)

27.1–28.31	The Journey to Rome and Earth's Ends. The drama of the voyage to come 'to Earth'. Paul's is preaching 'openly and unhindered'.

There are three editing idiosyncrasies to note in the pages that follow.

- Unless otherwise indicated, I offer my own translation of Acts. Because of my preference to err on the side of literalness rather than a dynamic translation, the text may appear a little stilted. Literalness helps to isolate specific linguistic or thematic details lost in more conventional translations. These are further assisted by identifying verse numbers, occasionally dividing these into smaller units (a, b, c . . .). Sometimes a transliterated Greek word will appear immediately after its English equivalent. This allows for an exploration of the original meaning of the word relevant to the present work.
- The intertextual approach I adopt in this ecological listening to Acts requires skimming over some very important stories, perhaps mentioning them in passing but not studying them in depth, though they deserve the greatest attention. This is, of course, the disadvantage of the present commentary. The redress to this is found in the exegetical interpretations offered in those fine commentaries mentioned earlier.[30] I would hope, though, that an overview of Luke's story in Acts would still emerge in the present contribution.
- Finally, the clear connection between Luke and Acts in terms of authorship and themes will mean frequent reference in the footnotes to *AEC* and the ecological thematic presented there. Summary environmental insights explicated in *AEC* will be presented in this volume where relevant. However, I invite the reader to consult *AEC* for their fuller treatment.

As I travelled through Turkey, Israel and Greece as this present work was gestating, I was encouraged through the gracious generosity and insights offered to me by local guides in these countries: Hakan Bashar in Turkey, Carol Ann Bernheim in Israel and Eva Vasillatou in Greece. Being with these wonderful people over several years and bringing students and pilgrims from Australia, I have appreciated more and more their professional expertise, and their care of those who accompany me to these countries. These are good friends. I thank each of them for their accompaniment to me as I shared with them, though they might not have realized it at the time, some of the insights contained herein.

I also want to acknowledge the careful reading of earlier drafts of this book by Rosemary Hocking. Her quick eye spotted many editing infelicities. Their correction and Rosemary's editorial skill have made this a better piece of writing than it would have been otherwise.

In the writing of *About Earth's Child*, my father, Laurie, died. I dedicated that volume to him. As I began to write this present volume my oldest brother, Terry, though relatively young, also died. Like our father, he had a desire to act in a way that was sensitive to the world around him. In his later years he sought to make his house more environmentally friendly. Though finances and pragmatics might have nudged him in this direction, he became convinced there were economic consequences in the way Earth's goods were used and people lived.

When I came into the final stages of completing the writing of this book, another 'brother' died. He was Denis Edwards, priest, theologian, friend and companion. He

---

[30] See the commentaries mentioned in footnote 7.

was a brother to me in many ways. We shared community for several years and he accompanied me in ministry, theological education and pastoral discernment. He was most encouraging of this present work. A prominent theme of his own theological writings was about ecology and retrieving the Christian tradition to offer theological depth for Christians involved in eco-theological work. His writings focused on the environment and the care of the natural world.[31] However, there was something else that I discovered in reading one of his books that lends further affirmation from him for this present volume.

Denis was inspired by the theology of Karl Rahner. In one of Denis' last writings he reflects on a brief article which Rahner wrote in 1950, entitled, 'A faith that loves the Earth'.[32] Here Rahner meditates upon the implications of Jesus' death and resurrection for the Earth. Rather than seeing Jesus' death as an escape from Earth into resurrected heavenly glory, Rahner believes that Jesus enters the very depth of Earth through his death, bringing God's own heart to it:

> In his death, the Lord descended into the lowest and deepest regions of what is visible. It is no longer a place of impermanence and death, because there he now is. By his death, he has become the heart of this earthly world, God's heart in the centre of the world, where the world even before its own unfolding in space and time taps into God's power and might.[33]

God's heart lies at the centre of the Earth through Jesus' death. His resurrection is not an abandonment of Earth and its creatures, 'but profoundly connected to all that is bodily' and all that is Earth-related.[34] Denis quotes Rahner again:

> No, he is risen in his body. That means: He has begun to transfigure this world unto himself; he has accepted this world forever; he has been born anew as a *child of this earth*, but of an earth that is transfigured, freed, unlimited, an earth that in him will last forever and is delivered from death and impermanence for good.[35]

Rahner's thought about creation and the meaning of bodily resurrection as refracted through Denis' own work is remarkably Lukan. Rahner considers Jesus as a 'child of

---

[31] Denis' works include: *Deep Incarnation: God's Redemptive Suffering with Creatures* (Maryknoll, NY: Orbis Books, 2019); *Christian Understandings of Creation: the Historical Trajectory* (Minneapolis: Fortress, 2017); *How God Acts: Creation, Redemption, and Special Divine Action* (Minneapolis: Fortress, 2010); *Jesus and the Natural World: Exploring a Christian Approach to Ecology* (John Garrett: Melbourne, 2012); *Ecology at the Heart of Faith* (Maryknoll, NY: Orbis Books, 2008); *Breath of Life: a Theology of the Creator Spirit* (Maryknoll, N.Y.: Orbis Books, 2004); *Jesus and the Cosmos* (Eugene, OR: Wipf & Stock, 2004); *The Natural World and God: Theological Explorations* (Hindmarsh: ATF Press, 2017); *Made from Stardust: Exploring the Place of Human Beings within Creation* (North Blackburn, Vic.: Collins Dove, 1992).
[32] Karl Rahner, 'A Faith that Loves the Earth,' *The Mystical Way in Everyday Life: Sermons, Essays and Prayers: Karl Rahner, S.J.*, ed. Annemarie S. Kidder (Maryknoll, NY: Orbis Books, 2010), pp. 52–8. The discussion on Rahner's article takes place in Edwards, *Christian Understandings*, pp. 222–4.
[33] Rahner, 'A Faith,' p. 55, as quoted in Edwards, *Christian Understandings*, p. 222.
[34] Edwards, *Christian Understandings*, p. 222.
[35] Rahner, 'A Faith,' p. 55, as quoted in Edwards, *Christian Understandings*, p. 223. Emphasis added.

this earth' being born 'anew'. Denis sees Rahner's interpretation as confirming that Jesus is at the heart of the yearning of all creation to participate with him in the transfiguration of his body. Denis concludes, 'Earth is our mother, and we are children of Earth, and we are called to love her.'[36] From a Rahnerian perspective, then, Jesus is *Earth's Child* and those attuned to him are *Earth's Children*. These two images overlay the particular ecological approach in listening to the narrative dynamic in Luke's gospel and the Acts of the Apostles. Luke-Acts is the story of *Earth's child* and *Earth's children*.

To Terry and Denis, two of Earth's children and, now, more-so in death, I dedicate this book, *About Earth's Children: An Ecological Listening to the Acts of the Apostles*.

---

[36] Edwards, *Christian Understandings*, p. 223.

Part One

1

Acts 1.1-5. An Ecological Orientation

The opening verses of Acts (1.1-5) are a prologue and summary.[1] Its author recalls what are essential memories from Luke's first volume, the gospel, and link to what is about to unfold in the evangelist's second major piece of writing. These verses also provide an ecological orientation for what will occur in the rest of Acts.

> In the first word, Theophilus, I dealt with everything which, in the beginning, Jesus did and taught up until the day, after having given commandment through the Holy Spirit to his apostles whom he had chosen, he was taken up.
>
> 1.1-2

Acts' Prologue (1.1-5)

Looking back, Luke offers a fresh summary of the gospel which he calls 'the first word' (1.1a). These verses concern the story of Jesus' words and deeds 'from the beginning' and conclude with his ascension after his spirit-inspired instructions to the apostles (Lk. 24.50-53). Then, in an expansion on the gospel story, which listeners would not have known until now, Luke tells how Jesus appeared for forty days after his death and resurrection speaking about God's 'reign' (*basileia* – and sometimes translated as 'kingdom') (1.3). In his final instruction he encourages his disciples to remain in Jerusalem, to await the Father's promise of baptism by the Holy Spirit (1.4-5). Finally, note that these introductory verses are addressed to Theophilus (1.1a), the same addressee of Luke's gospel (Lk. 1.3).

The mention of Theophilus and the designation which Luke gives him in the gospel's prologue (Lk. 1.3) as 'most excellent' indicates a person of elite social status. The presumption is that this is the same character addressed here in Acts and, perhaps, to whom Luke dedicates the volume. Though the gospel (and now Acts) is addressed to him explicitly, it is not intended for one person, but an audience of Jesus followers in the late-first-century Greco-Roman world.[2] This might also be supported by the

---

[1] Luke draws on similar prologues in the Hellenistic world. See Pieter van der Horst, 'Hellenistic Parallels to the Acts of the Apostles: 1.1–26,' *ZNW* 74 (1983), pp. 17–26.
[2] See the discussion on Theophilus in *AEC*, pp. 7, 24–9, 38, 49.

thought that 'Theophilus' might be a symbolic designation as it refers to one who is a 'lover of God', presumably all within the Lukan household.

While perhaps Luke might have in mind a group of elite Jesus followers – and it is to these that Acts, like the gospel, is addressed – Acts is not exclusive of other social and cultural groups present in Luke's audience. As indicated in the previous chapter, they would have various attitudes to Earth's gifts.[3] Some members of the Lukan community would identify with Theophilus, given a similar social status and their collateral relationship with him. These, as in the gospel, would also be the primary addressees of Acts. They would influence the social context in which Luke pens this second volume written in a Greco-Roman world of social status and hierarchy. The growth of the Jesus movement beyond Judaism, Jerusalem and Judea would be dependent on their support, their ability to let go of a status dependent on wealth and influence, allow their wealth to be available for the use of the wider Jesus gathering and its mission (this is the background for Luke's story of Ananias and Sapphira in 5.1-11), and join in communion with all those members of the Jesus movement from other social contexts and on different rungs of the Greco-Roman social ladder (Figure 1). Issues of wealth, social communion and the encouragement to move beyond what is familiar and comfortable become a strong focus in the early chapters of Acts. These reach their highpoint at Act's narrative midway point when the Jerusalem leaders need to consider the future direction of the Jesus movement beyond Judaism and its potential embrace of the non-Jewish world (Acts 15).

These opening verses situate what is about to take place in the context of Luke's story of Jesus. They also anticipate the next moment in the story that will unfold with the earthly Jesus' return to the Father (1.9-11) and the coming of the Spirit upon Jesus' chosen ones at Pentecost (2.1-13). Luke looks back and looks forward.

At a first hearing, these verses concern Jesus and the disciples who will appear as central characters in Acts. At a deeper level of listening, at an ecological level characterized by the mode of intertextual listening summarized in the previous chapter, these opening verses offer the beginning of an environmental orientation that continue in Acts. They offer reminders of Luke's central gospel truths that have ecological resonances often shrouded by an anthropocentric focus.

Luke's Gospel Remembered

As mentioned earlier in the summary of Luke, the gospel concerns Jesus. He is Earth's Child. He acts with sensitivity to all creation, including humanity. Here, in these opening verses of Acts, the contemporary audience, approaching Luke's story with an ecological hermeneutic, hears other undertones, ecological memories from the gospel that need recalling. These establish listening perspectives that will guide the present commentary as I follow how Jesus' disciples, Earth's Children, become agents of his message to the whole Earth. There are four that are key.

---

[3] For a hypothetical construction of the social composition of the Lukan audience, see *AEC*, pp. 26–30.

The 'Word'

First, Luke describes the earlier writing, the gospel, as a 'first word' (*protos logos*). This is often translated as 'first book' (NRSV) and there are Hellenistic precedents to confirm the author's thinking in this way.[4] Luke is referring to the gospel as a 'first word' or 'book'. This suggests to Theophilus that the present is a 'second' word or book. However, at another level of inference, 'word' has other connections in the gospel. 'Word' is a primary linguistic root-metaphor for communication. Without it, we would be locked into a world of silence, solitariness and disengaged vitality. In the ancient and Lukan thought-world, 'word' is a dynamic reality that brings about meaning, change and life. For Luke, the communicator of this 'word' is Jesus who acts, enacts and enables its receptor, the listener, to come to a deeper sense of meaning and encounter with the sacred. The receptor of Jesus' word, though, is not only human. As seen in the Lukan study, Jesus speaks to and about human and non-human beings. His word *effects* creation. He speaks with authority, which surprises his audience and his disciples. His word also heals.

The 'Word' creates. So rather than only being a defined cognitive or noetic element of human communication, Luke's 'word' also acts.[5] It has a powerful agential dimension that brings about change, communion and healing. Luke's description of the gospel as the 'first word' refers to the whole story of Jesus from the 'beginning', to the gospel's last scene when Jesus ascends to God on the day of Easter (Lk. 24.50-53). Luke refers to this 'beginning' explicitly in the gospel's prologue, acknowledging that the gospel narrative is the fruit of the evangelist's 'following everything closely from the very beginning' (Lk. 1.3).

For ears attuned to the Torah, the 'beginning' thematically echoes the early chapters of Genesis and its stories of creation. These are the Torah's 'beginning'. Luke's 'beginning' may not be controlled only by the logical narrative commencement of Jesus' 'beginning', his birth and his early moments in his Galilean ministry. It might also embrace all the elements of creation implied in 'the beginning' from Genesis 1.1.

In whatever way the contemporary listener hears 'beginning', its repetition in the opening verse of Acts centres on the 'beginning' of Luke's story of Jesus that slowly reveals the gospel's Christology. This story is about God's Spirit-filled agent who is in love with Earth and its creatures, and seeks to bring them into communion with God. Luke's implied 'second word', the *Book of Acts*, will demonstrate how this 'first word' finds its continuity in the story of Jesus' followers and in their engagement with Earth and its inhabitants.

God's 'Reign' as '*Basileia-Ecotopia*'

Second, as Luke summarizes the 'first word', the evangelist anticipates a second ascension scene which will be recounted soon. The author tells listeners that Jesus appeared forty days after his passion with various proofs speaking about the 'reign

---

4   For example, Philo, *Prob* (= *Quod Omnis Probus Liber Sit*) 1.
5   On the meaning of 'word' (*Dabar/rhēma*) in Luke see *AEC*, pp. 73–5.

(*basileia*) of God' (1.3c). If Jesus has already ascended once (in Lk. 24.50-53) in an event on Easter day that consummates Jesus' resurrection, it is clear now that his return to the Father will also take place forty days later as a forecast of the Christological presence in Acts. The Acts ascension clearly offers a constant witness to the way Jesus will manifest himself throughout Luke's 'second word'. Essential for what follows is the emphasis which Luke places on Jesus' final Earth-centred communication in terms of the 'reign of God'.

The gospel reveals how the 'reign' manifests itself. It is a central theological reality for Luke who mentions it in the gospel over forty times. Luke presents Jesus' revelation of God's *basileia* as an inclusive encounter with the divine presence culturally, socially and *ecologically* manifest in the particularity of the world in which Jesus ministers.[6] The *basileia* is not a hierarchically derived entity, mirroring a universal heavenly scale that is obvious within the ancient Greco-Roman world of social structure. Rather, it is an inclusive cosmic symbol that prepares, reveals and anticipates the eschatologically oriented desire of God's communion with all creation. The speech and action of Luke's Jesus concern the revelation of God's *basileia* as essentially inclusive, healing, reconciling and uniting. It brings freedom and counteracts any expression of evil that antagonizes or destroys creation. As noted in the Introduction, God's *basileia* is also an '*ecotopia*'. It anticipates and realizes God's intent for all creation, including humanity.

Luke underscores the nature of this *basileia-ecotopia* in several places in the gospel (for example, Lk. 4.42-44; 5.12-32; 7.31-32; 8.4-39; 10.8-9, 16; 13.18, 21–22, 29; 17.20-27, etc.). Its presence continues in Acts. Its memory characterizes Jesus' post-resurrection forty-day preaching (1.3c) that encourages his disciples to remain in Jerusalem and to 'wait for the Father's promise which you heard from me, that John baptized with water, but shall be baptized with the Holy Spirit before many days' (1.4b-5).

'Water' and the 'Holy Spirit'

Third, Acts 1.4b-5 completes Luke's prologue to Acts. It recapitulates other gospel themes and images that have ecological inferences. Two of these are 'water' and the 'Holy Spirit', first explicated in the baptismal ministry of John the baptizer ('the dipper'[7]) where they are linked.

Water

Water, the primeval element of creation, is implicit in the gospel's opening scene as the angel approaches the priest Zechariah in the temple (Lk. 1.5-23). This is the symbolic navel of the universe where all Earth's fruits are symbolized and Genesis' watery origins celebrate in the four large bronze basins (called 'the sea') that adorn the courtyard entrance to the temple.[8] Water is also implicit in the scenes that surround the meeting

---

[6] For example, Luke 4.42-4; see *AEC*, pp. 129–31.
[7] Joan Cecilia Campbell, *Phoebe: Patron and Emissary* (Collegeville, MN: The Liturgical Press, 2009), p. 8.
[8] On the environmental richness of the temple and its 'water', see *AEC*, pp. 69–73.

of the pregnant Mary and Elizabeth whose birth waters carry the children they bear (Lk. 1.39-45) in imagery that is gestationally powerful.

In Luke, water also accompanies the presence of the Spirit, God's presence, and permeates the gospel's opening scenes and saturates images of fecundity and fertility. The pregnancies of Mary and Elizabeth, the fruits of the Spirit's action, are deep ecological images that promise a vivifying future. They anticipate the divine intention for Earth's ongoing fecundity. In the body of the gospel, water features again as Jesus sails upon it on a boat, restoring it to its divinely designated calming order after it is lashed by demonic forces that have invaded Earth's elements (Lk. 8.22-25).

Water also features in Acts. It is explicit in several important baptismal scenes which occur at significant moments as the Jesus movement spreads beyond Jerusalem, the place of its origin. These baptismal events forecast the next stage in the movement: Stephen baptizes the Ethiopian eunuch (8.26-40); Peter uses it in baptizing Cornelius' household (10.48); Paul's act to Lydia and her household at Philippi, the first 'conversion' on Macedonian soil (16.15). Water features at key strategic narrative moments in Acts.

Besides its presence in these stories, water is the most prominent though usually unnoticed Earth image that helps sustains Paul's journeys. He sails upon it. He boards various shipping vessels, themselves the constructions from Earth's materials (a point which I take up in a later chapter) to carry him over the waters on all four main voyages:

- From Antioch to the coast of Pamphylia in Asia Minor via Cyprus (13.4-13), back to Antioch from the southern Asia Minor port of Attalia (14.26).
- From Troas on the western Anatolian coast across the Aegean to Neapolis in Macedonia (16.11).
- From Corinth across the Aegean and Mediterranean seas to Syria, Caesarea Maritime and Antioch (18.18-22); from Philippi (20.6), Assos, Mitylene, Chios, Samos, Miletus (20.13-15), Cos, Rhodes, Patara, Tyre in Phoenicia (21.1-4), Ptolemais, to Caesarea (21.8).
- Then a final voyage from Caesarea to the Roman port of Puteoli (27.2–28.13). This last voyage of Paul is so detailed and dramatic that it will be the main focus of Chapter 13.

More will be said of water's ecological symbolism when I look in greater detail at these various sections of Acts concerned with Paul's journeys. For now it is helpful to note the role which water and its seas play in Luke's story of Paul. It is an ecological backdrop.

The Holy Spirit

In the gospel and in the ministry of John, the evangelist creates a relationship between water and the Holy Spirit. This link symbolically parallels a deeper theological connection: that between Heaven and Earth.

After the gospel's birth narrative (Lk. 1–2), the author introduces John and his mission of social reform (3.1-14). In response to the expectation that he might be the messiah (3.15), John identifies himself as the one who baptizes with water, but then declares that one mightier than he will come who 'will baptize you with the Holy Spirit and fire' (3.16). The connection between baptism and the Holy Spirit is graphically narrated later in the scene of Jesus' baptism (3.21-22). Luke's redaction of the same scene borrowed from Mark (Mk 1.9-12) clearly shows that Jesus' baptism is not the work of John whom Herod incarcerates in the next scene (Lk. 3.18-20) and before Jesus is baptized. His baptism is thus God's action. In Luke's reconstructed baptismal prayer scene, Jesus' baptism is a moment of prayer, an encounter with God, in which 'Heaven was opened, and the Holy Spirit descended upon him in bodily form like a dove and a voice came from Heaven "You are my Son, the Beloved; with you I am well pleased."' (Lk. 3.21c).

The memory of this event is recalled by Luke at the beginning of Acts. There are several aspects about this for listening to Acts. The opening of the Heavens and the descent of God's Spirit upon Jesus are Luke's narrative symbols that convey an important Christological truth: Jesus is the bearer of God's Spirit that is no longer kept in the Heavens, but is now accessible to all who encounter Jesus. This divine accessibility imaged through Heaven's 'opening up' is a reminder that the divine realm is not a closed system to which only the holiest have access, and then after death. It is a present reality available to all. The heavenly realm of God's presence is now tangible within Earth, explicated in Luke's scene of Jesus' baptism and anticipated earlier in events that surround his birth.

In Luke's story of Jesus' birth (Lk. 1–2) several Earth elements are central to the narrative. The evangelist twice mentions that Jesus is 'wrapped in bands of cloth' (2.7,12) and three times placed in a manger after birth (2.7, 12, 16). Both have ecological implications.[9] The cloth that surrounds the child and the manger which holds him are gifts of Earth and theological images emphasized in Luke's story. The cloth wrapping identifies who Jesus is: we are what we wear. He is a child of Earth. The stone receptacle in which he is laid is associated with food and animals. These associations further deepen the ecological connection which Luke makes in the narration of Jesus' birth. They also orient the gospel listener to the ongoing ecological and Earth connectedness which Jesus will have with all Earth's creatures, human and non-human, in the gospel's subsequent chapters.

Communion between Earth and Heaven

As God's Spirit descends from the Heavens now 'opened' through Jesus' prayer at his baptism, the communion of God with Earth, of the heavenly realm with the earthly, anticipated in the narrative symbols (cloth wrapping and manger) of Jesus' birth, becomes more firmly explicated in the angelic hymn sung to the shepherds who witness their own divine encounter (Lk. 2.14). The hymn is carefully constructed in such a way to illustrate that the apparent separation of the two cosmic spheres (Heaven and Earth) now collapses in the birth of Earth's Child who brings peace, the fruit of divine communion, among members of the Earthly community. Heaven, as it were, comes to Earth. Jesus' birth enables human beings, creatures of Earth, to realize their blessedness (*eudokia*):

---

[9] *AEC*, pp. 79–85.

```
        Glory (doxa)
            among the highest (hupsistos) to God,
               and upon Earth, peace
            among human beings
    of good will / favoured / chosen (eudokia)
                                                    Lk. 2.14
```

Figure 3 Literary pattern of Lk. 2.14.

The hymn is crafted with balance, centred on 'and upon Earth, peace' (Figure 3). Peace is the result of Jesus' birth and his communion with creation. Heaven comes to Earth. The hymn's resolution of the spherical separation between the realms of Heaven and Earth celebrated in the birth of Earth's Child has its echo in the beginning of Luke's story of Jesus passion (Figure 4), as he enters Jerusalem (Lk. 19.38). As his disciples welcome Jesus on Jerusalem's outskirts, they sing:

```
    Blessed is the one who comes – the King – in the name of the Lord
               in Heaven, peace
    and glory (doxa) in the highest (hupsistos)
                                                    Lk. 19.38
```

Figure 4 Literary pattern of Lk. 19.38.

Before the gospel's denouement and Jesus' death takes place, this hymn sums up his public ministry. Elements from the birth-hymn recur: the presence of 'glory' (*doxa*); the celebration of the 'highest' (*hupsistos*: the heavenly realm associated with God's abode, a point taken up in the next chapter), the acknowledgement of one of Earth's creatures ('the blessed kingly one'), and the gift of peace, the centrepiece of both hymns. In the birth-hymn (Lk. 2.14), Heaven comes to Earth and brings peace to Earth. In the hymn welcoming Jesus into Jerusalem (Lk. 19.38), Earth comes to Heaven and brings peace to Heaven.

By the end of the gospel and because of Jesus' ministry, the ministry of Earth's Child and Spirit bearer, Earth and Heaven experience peace. Throughout the gospel narration of Jesus, from his birth to the final days of his ministry, Heaven and Earth merge. There is no longer any separation; the divine infuses Earth and Earth pervades Heaven. Earth thus becomes the authentic revealer of the Sacred. Luke's story of the growth of the Jesus movement in Acts will be played out against Earth's backdrop and this conviction.

Conclusion

These first five verses that open the Book of Acts look back and look forward. They summarize for the auditor Luke's gospel centred on Jesus' deeds and words (1.1) covered by the 'beginning' of the story and conclude when Jesus was 'taken up'. Jesus' birth and his ascension, the gospel's final story, frame Luke's 'first word'. These verses also remind the audience of Jesus' apostolic commission in the gospel's last chapter. This commission, here described as 'commandment through the Holy Spirit', is one of 'reconciliation for the forgiveness of sin' (Lk. 24.47).

At one level these words could be heard as a call for the apostolic community to engage in a mission of reconciliation amongst peoples of different cultural and social backgrounds to bring about peace, 'the forgiveness of sins' and, in the economic convictions that shape Luke's writings, to share their possessions and Earth's goods with a humble spirit of ecological asceticism. The ecological listener could hear Acts 1.1-5 in another key: as an invitation to bring reconciliation within the Earth community, of all God's creation, in the spirit of Luke's Jesus. This Earth-oriented ecological mission of reconciliation brings communion within the Earth household, human and non-human. This will result in the 'forgiveness of sin', that is, the release of any form of brokenness that holds Earth to ransom or destroys the network of communion divinely intended.

The Acts prologue invites the contemporary listener to remember ecological themes and images that link to the gospel's portrait of Jesus: his Earth-connectedness underscored first in the image of bands of cloth and manger associated with his birth; then in his table communion with Earth's forgotten ones; his injunction to his disciples in their use of wealth and Earth's goods; and his healing and teaching ministry concerned about human beings, animals, wind and water. Jesus reveals God's presence in his teaching about the *basileia* which is also an *ecotopia*. All creation is the focus of Jesus' ministry.

Finally, this return to the insights from Luke's gospel is not a distraction from what is important in these opening verses in Acts. Emphatic is the recognition of the divinely ordained communion between Earth and Heaven celebrated in the gospel through the ministry of Jesus and the ecological focus of a ministry not exclusively anthropocentric. This conviction of the communion between Heaven and Earth will have its continuity in the role which Jesus' disciples play as Luke's story of Acts now unfolds.

2

Acts 1.6-11. The Ecological Mission

As the Child of Earth inaugurates the *basileia-ecotopia* in the gospel, Jesus' disciples and those empowered by the Spirit, Earth's Children, will continue this mission in Acts. The scene immediately following the opening verses of Acts (1.6-11) offers a final directive from the Risen and Earthly Jesus. His words confirm the programme which will unfold in the rest of Acts. They reveal the direction which Luke will take in developing the story of Earth's Children.

Jesus' Final Words to his Disciples (1.6-8)

The disciples seek to identify the exact manner and chronology of the *basileia-ecotopia*'s manifestation to bring about Israel's restoration. They ask, 'Lord is this the time when you will restore the *basileia* to Israel?' (1.6). Jesus reminds them that God's authority determines everything (1.7). Jesus' response aligns with the ecological insight gleaned earlier in our consideration of the particularity of Luke's presentation of the *basileia-ecotopia*, revealed in Jesus' ministry. 'Jesus' answer contains a redefinition of "kingdom",' writes Luke Timothy Johnson, 'and therefore of the Christian understanding of Jesus as Messiah. The kingdom of God is not a territory or political realm. It is the rule of God over human hearts'.[1] And it is more. It is an encounter with God's presence (one might say God's glory: *doxa*) intended for the healing (peace: *shalom*) of all creation, human and non-human.

The disciples' question to the Risen Jesus at this early stage in Acts seeks to determine and, to a certain extent, control the *basileia*'s encounter. Luke's Jesus reminds his listeners that its mystery remains God's initiative, not frustrated by human designs. It is already present 'among them'. This God that Jesus acknowledges is the One whose authority fixes 'times and seasons'. The image of God which Jesus presents in response to the disciples' query is the God of Creation. Jesus' gentle rebuke of his questioners redefines the conventional expectation of what the *basileia-ecotopia* looks like, and moves them to consider their mission. Jesus' final words are clear: 'But you shall receive power (*dunamis*) when the Holy Spirit has come upon you, and you shall be my

---

[1] Johnson, *Acts*, p. 29.

witnesses (*martus*) in Jerusalem and in the whole of Judea and Samaria and to the end of the Earth (*gē*).' (1.8). This is an important verse. For many commentators it establishes the narrative division in Luke's long story of Earth's children. They see a three-fold division that centres on Jerusalem (1.1–8.3), then what occurs in Judea and Samaria (8.4–15.35), finally, in the longest section of Acts (15.36–28.31), Paul's three journeys, that eventually bring him on a final journey to Rome, Earth's thought 'end', where Acts concludes. Their consideration is geographically determined by Acts 1.8c. Earth's 'end' is not really an end, at least in terms of geography, as seen later.

From another perspective, one that has been advocated by Edward Soja, it is possible to view Jesus' geographical markers as 'spatial'. This has ecological implications. Soja critiques a process of narrative analysis primarily historical; this can apply to a biblical interpretation of Acts that is essentially historical. This approach, according to Soja, comes from western social theory that emphasizes 'a superficial linguistic spatialization that makes geography appear to matter theoretically as much as history'.[2] However, it is possible to hear Jesus' injunction as a consideration of the geographical spatialization implied behind the topographical markers, Jerusalem, Judea, Samaria and Earth's ends (1.8c). This makes the geography that shapes Luke's story more than a determination of cultural or ethnic identity with which Earth's children engage. Geography becomes, 'dynamic and pluriform, sensitive to heterogeneity, conscious of the politics of presentation and interpretation, and eager to allow room for new and marginalized ways of writing the earth. It is a multi-headed, interdisciplinary – even 'post-disciplinary' – creature'.[3]

For the auditor of Acts, Jesus' spatial markers are not geographical containers that define and control Luke's narrative of Earth's children. Rather they reflect social and spatial networks that indicate and configure ecological beings and environmental relationships in the Greco-Roman world.[4] This is the nature of space considered ecologically.

There are three other elements to Jesus' injunction that complement this geographical division and spatial consideration. These have clear ecological resonances and concern (1) the promise of 'power' through the action of the Holy Spirit, (2) the urge to witness and the nature of the implied witness – especially given the description of discipleship witness that the careful biblical listener can remember from Jesus' teaching in the gospel's

---

[2] Edward W. Soja, *Postmodern Geographies: the Reassertion of Space in Critical Social Theory* (London: Verso, 1989), p. 7, as quoted by Matthew Sleeman, *Geography and the Ascension Narrative in Acts* (Cambridge, UK: Cambridge University Press, 2009), p. 23. Soja distinguishes between three 'spaces': 'firstspace' represents concrete material forms that can be controlled and analyzed; 'secondspace' is the intellectual or cognitive representation of space and its social analysis; 'thirdspace' lies beyond this, offering other forms of spatial representation and idealization. It is in this 'thirdspace' that an ecological alertness to the spatial implications of Acts belongs. For further reading on 'thirdspace', see Edward W. Soja, *Thirdspace: Journeys to Los Angeles and Other Real-and-Imagined Places* (Hoboken, NJ: Wiley, 1996); Karin Ikas and Gerhard Wagner (eds), *Communicating in the Third Space* (Routledge Research in Cultural and Media Studies No. 18, London: Taylor & Francis, 2009).

[3] Sleeman, *Geography*, p. 25.

[4] Joel B. Green, 'The Death of Jesus and the Rending of the Temple Veil (Luke 23:44–49): a Window into Luke's Understanding of Jesus and the Temple,' *The Society of Biblical Literature 1991 Seminar Papers*, ed. Eugene H. Lovering Jr (Atlanta, GA: Scholars Press, 1991), pp. 543–57.

journey narrative (Lk. 9.51-19.27) – and (3) the specific locations (Jerusalem, Judea and Samaria) concluding with 'Earth's end', the culmination of the disciple's mission.

'Power' (1.8a)

First, Jesus' promise to his disciples, that they 'will receive power when the Holy Spirit' comes upon them (1.8), looks forward to the Pentecost event which the author describes in the next chapter. This Lukan technique, of anticipating or flagging in an earlier scene what will occur later, provides the key elements for which the listener is attentive. This anticipatory technique looks forward; it also looks back. Mary, the primary disciple, is Luke's example. In her encounter with God's angelic presence in the gospel's first chapter (Lk. 1.26-38), perplexed by the angel's greeting and more confused by the fact that she will give birth, she hears God's word: 'The Holy Spirit will come upon you, and the power (*dunamis*) of the Most High will overshadow you; therefore indeed the begotten one will be called holy, the Son of God.' (Lk. 1.35).

According to the angelic voice, Mary's reception of God's Spirit is accompanied by her experience of God's power (*dunamis*) through the presence of the Spirit which 'overshadows' her. This dynamic, of the Spirit and the associated *dunamis* that accompanies the Spirit's presence, initiates Mary's fecundity and confirms the holiness of the one whom she carries in her womb. Her openness to the Spirit which 'overshadows' her (Lk. 1.35b) allows her to experience the vivifying agency of God who will bring to birth God's promise in Jesus, Earth's Child. His birth, the gift of the Spirit, and future ministry, ensured by the Spirit's presence and promise that comes upon him in prayer at his baptism (Lk. 3.21-22), have ecological implications. He is identified with Earth, surrounded by Earth's gifts at birth, and acts in word and deed in his public ministry to bring healing to Earth's inhabitants, human and non-human. God's Spirit thus acts with ecological generativity.

This same interrelationship – Spirit, power, holiness, fecundity – observed in the gospel, is remembered by Luke's audience in the words which the Risen Jesus speaks to his gathered disciples before he departs from them. It foregrounds the Pentecost event in Acts 2, as the Spirit descends on the disciples in the Jerusalem upper room.

The *dunamis* acting in the gospel overshadowing Mary and promised to the gathered disciples in the beginning of Acts is a favourite Lukan expression.[5] It has a rich association in the Greek world. The Greeks considered it a dynamic cosmic principle which Plato identifies with the essence of Being and associates it with God.[6] It is periphrastic for God's presence: essentially creative.[7] God's *dunamis* is concerned

---

[5] *Dunamis* occurs 119 times in the NT, almost a quarter in the Lukan corpus (15 in the gospel; 10 in Acts). Luke inserts *dunamis* four times into the Markan source upon which Luke draws, and four times into material that is unique to the gospel. See also Max Turner, *Power from on High: the Spirit in Israel's Restoration and Witness in Luke-Acts* (Sheffield: Sheffield Academic Press, 2000), esp. pp. 428-38; Robert Menzies, 'Spirit and Power in Luke-Acts: a Response To Max Turner,' *JSNT* 15 (1993), pp. 11-20.

[6] Pl., *Cra.* 404e; 406a.

[7] *Dunamis* as a characteristic of God is evident in Greek and OT literature. For example, Pl., *Cra.* 406a; Josh. 4.24; Deut. 3.24; LXX Pss. 76.15, 144.12; Jer. 16.21; Jdt. 9.8, 13.4. See also G. Friedrich, 'Dunamis,' *EDNT* 1, (Grand Rapids, MI: William B. Eerdmans Publishing Company, 1990), pp. 355-8.

with the prolificacy of all creation. This same fecund Spirit is promised to the disciples, soon to be poured out over them.

'Witnesses' (1.8b)

Second, Luke's Jesus tells the disciples that they are going to be his 'witnesses' (*martus*) (1.8b).[8] This apostolic designation adds further insight into the kind of witnesses that Jesus intends them to become. They are *his* witnesses. This Christological frame of reference shapes their future ministry that will unfold in the rest of Acts. This reference implies two things: (1) They are not to be private, self-centred purveyors of what they think they are called to be and do. Their witness is about Jesus and his ministry as revealed in Luke's gospel. (2) Their witness is not solitary or isolated. It is Christologically linked. As Beutler suggests, 'Luke develops a usage according to which the apostles are "witnesses" not only to the outward events of Jesus' life, death, and resurrection, but also to their salvific meaning according to Scripture . . . [*Martus*] expresses not only the element of valuation but also that of *personal engagement*.'[9]

Whatever they do and say, their bond with the Risen Jesus remains and, through their witness, the Risen One will also be present, albeit not directly as he appears in the concluding chapter of the gospel or in the first chapter of Acts before his ascension. His presence will be through his faithful witnesses. They have encountered the Risen Jesus, they are witnesses to his death and resurrection, and this has touched their lives. Their witness is personal and intentional.

The disciples are now explicitly entrusted to this witnessing characteristic that will define their ministry in Acts. They are to witness the person of Jesus, echo his ministry, his deeds and actions, in what they do and say. This connects to the promise and injunction of the Risen Jesus on Easter day in the gospel's final chapter. Jesus appears to his disciples on the day of his resurrection. He eats with them (Lk. 24.41-42). This reassures them of the reality of his presence, that he is not pure spirit (Lk. 24.39). The words he speaks that accompany this meal contextualize the biblical memory for Luke's audience. As the Risen Jesus 'opens the minds' of his disciples to the Scriptures (Lk. 24.45) he reminds them that the events that have occurred, especially his passion and resurrection, are in harmony with the Jewish biblical traditions found in the Torah, Prophets, and 'the Psalms'.[10] He reminds them that their future preaching of repentance and forgiveness of sins is grounded in this biblical heritage. Their primary status in the future is as witness to this message that Jesus now announces. Repentance and forgiveness flow from this. This 'repentance for the forgiveness of sin', as seen in the previous chapter, concerns the disciples' Earth-related mission that brings *shalom* to the total Earth community. The location of this mission is explicated in the third part of our discussion on Jesus' final apostolic injunction.

---

[8] The expression 'witness' (*martus*) occurs most frequently in Acts (13 times). Like 'power', it is one of Luke's most important and favourite expressions.
[9] J. Beutler, 'μάρτυς', *EDNT* 2, pp. 393–5. Emphasis added.
[10] This is the classic threefold division of the Jewish Bible, in terms of *Torah* (the first five books of the OT), the prophets (the *Nebiim*), and the writings represented by the mention of Psalms (*Kethubim*), which gives the reference 'Tanakh' to the Jewish Bible.

'To the End of the Earth' (1.8d)

Luke has a clearly defined geographical region in mind: the area associated with Judaism (Jerusalem and Judea), with the description of the Jesus movement beyond Judaism first experienced amongst the Samaritans and then the expression 'to the end of the Earth' (1.8d). For some scholars, this designates the apostolic mission beyond to the whole Greco-Roman world, into Asia-Minor, Macedonia and Greece, and, finally, Rome.[11] The final designation is also an echo of Isa. 49.6. In what is recognized as Isaiah's Second Servant Song, God speaks to the prophet indicating that God's salvation is not reserved or limited to a particular people or place. It will take on universal proportions: 'It is too light a thing that you should be my servant to raise up the tribes of Jacob and to restore the survivors of Israel; I will give you as a light to the nations, that my salvation may reach to the end of the Earth.' (Isa. 49.6; NRSV).

Within the context of Isaiah in which this quote occurs, the whole song is not anthropocentrically focused or exclusive. It begins first addressing the natural world ('coastlands') before it moves on to address humanity ('Listen to me, O coastlands, pay attention, you peoples from far away!': Isa. 49:1; NRSV). The postscript which immediately follows celebrates the redemptive power of a God who liberates humanity and creation (Isa. 49.7-26). It is within this wider ecological setting that God's salvation reaches to 'the end of the Earth'. More is implied, then, than what will happen to humanity. Luke will again use Isa. 49 to conclude Paul's speech to potential Gentile converts in Pisidian Antioch (13.47).

In the classical world, Earth's 'end' had a specific cartographic distinctiveness in a cosmos which consisted of an inner land mass surrounded by an outer ocean ('*okeanos*'). The 'end of the Earth' was contiguous to the oceans.[12] It was where the inhabited and civilized world stopped. This is clear from the geographer Strabo (64/63 BCE–24 CE) who writes that the 'inhabited world is an island. For wherever it has been possible for humans to reach the end of the Earth, sea has been found. And this sea we call "Oceanus"'.[13]

The expression 'the end of the Earth' is a geographic identifier that locates the link between land and sea. It is also *Earth*. In the ears of the contemporary listener of Acts concerned about what happens on this planet, Earth represents more than a land mass linked to the oceans. It is an ecological image of interconnected relationships, a network of organic and non-organic matter and a living organism. From this perspective, Earth is a symbol not determined by human beings, but an interconnected vital system in which humanity lives and expresses its civilized reality. This recognition of 'the end of the Earth' as more than a specific geographic space not controlled by human beings or their presence, but the boundaries beyond and to the end of Earth's space, already

---

[11] The Rome association with 'Earth's end' comes (1) from the geographic termination of Acts in Rome and (2) the reference to Pompey's advance on Jerusalem in 63 BCE found in LXX *Pss. Sol.* 8.15: '[God] brought someone from the end of the Earth...He decreed war against Jerusalem.' Translation from E. Earle Ellis, '"The End of the Earth" (Acts 1:8),' *BBR* 1 (1991), p. 124.

[12] Willem C. van Unnik, 'Der Ausdruck "ΕΩΣ ΕΣΞΤΟΥΤΗΣ ΤΗΣ" (Apostelgeschichte I 8)...', *Sparsa Collecta, Volume 1*, eds J. Reiling, G. Mussies, P. W. van der Horst, L. W. Nijendijk (Leiden: Brill, 1973-83), pp. 386-91, as referenced in Ellis, 'The End,' p. 126.

[13] Strabo, *Geographica*, 1.1.8; cf. 1.2.31; 1.4.6; see Ellis, 'The End,' p. 126.

opens us to consider the 'more than' human meaning behind 'the end of the Earth'. It is beyond the human.

To venture to the 'end of the Earth' is to come to the unknown; it is fraught with danger, but also possibility. This 'end' can become a space where the new might be revealed in an eschatological healing of what has occurred through Earth's human habitation and abuse of Earth's goods. In fact 'end of the Earth' might be symbolic of several themes: the end of the inhabited world; the eschatological climax of God's vision for creation; the end of the Earth itself caused by the destructive forces initiated by human beings. 'End of the Earth' could be, for the Lukan contemporary listener, a multivalent symbol of creative possibilities or cautious action for the health of the planet. Jesus' final words to his disciples, to 'the end of the Earth', can also be for the contemporary disciple an ecological mission linked to Earth's outermost and fragile limits rather than a specific location.[14]

With this multivalence in mind, I move forward to consider how the disciples' ecological mission might unfold on Earth as they journey around a world circumscribed by 'Oceanus': the Mediterranean Sea. The disciples' relationship to Earth and its Sea is the environmental context for Luke's story, at times explicit, but generally implicit. As the disciples, particularly Paul, transverse Earth, its land mass and oceans are inherent in their mission.

Jesus Ascends (1.9-11)

This Earth, as the next scene underscores, is linked to God's abode through the ascension of Earth's Child into the heavenly abode. Their mission to Earth's end is endorsed by the Risen and ascended Jesus to whom they witness. His presence with God symbolized by his heavenly ascent is a bodily or physical enactment of the angelic birth hymn in an early chapter of Luke's gospel (Lk. 2.14) and the hymn sung by the disciples welcoming Jesus into Jerusalem towards the end of the gospel (19.38). Jesus' ascension in Acts reflects the communion of Earth with Heaven and the ongoing divine presence on Earth. The listener is again reminded that the cosmic spheres once differentiated, of Earth and Heaven, are one. The separation between the divine and the secular no longer exists, as Earth's Child lives eternally in the heavenly abode, and the sacred communion with what might be thought of as secular continues on the Earth through the ministry and mission of Earth's children.

Luke explicitly links the last words of the Risen Jesus, earlier explored, to this final act. As he speaks, he ascends. The two, the words and the actions, complement each other, the first continuing into the second. This means that the dominical injunction given to the disciples ('you shall be my witnesses in Jerusalem and in the whole of Judea and Samaria and to the end of the Earth [gē]') in 1.8 remains, even though the Earth-bound Jesus is no longer present in their physical sight. A cloud is central to the event. It is reminiscent of the same ecologically significant penumbra symbol of Jesus'

---

[14] Sleeman, *Geography*, p. 71.

holy encounter with God in his transfiguring event in Lk. 9.34-35.[15] In that event, God's voice speaks *from* the cloud. Here in Acts, the cloud becomes an agent of divine action, and 'takes' Jesus from the disciples' physical sight into the heavenly abode.

The communion between Earth and Heaven continues, as the disciples, while gazing into the Heavens, are reminded by 'two men' – the same two who appeared in Jesus' transfiguration in Lk. 9.30 and at the tomb in Lk. 24.4-5 – that Jesus will return 'in the same manner that you have beheld him go into Heaven' (1.11b). The return of Earth's Child is assured. The manner of this return will involve the ecologically symbolic element of nature, the cloud, that echoes Daniel's vision (Dan. 7.13) and the promised return of the Human One ('Son of Man'). The two men question the disciples who keep staring towards Heaven. Their focus is now elsewhere as they return to Jerusalem and await the promised Spirit who will empower them for their future mission and witness.

Jesus' ascension, his enthronement with God, completes Luke's introduction to Acts, defines the mission and witness that the disciples will undertake and, for the listener sensitive to the environmental issues that she encounters, Earth's involvement in this mission.[16]

Conclusion

There are important insights that emerge in the final words of the Risen Jesus to his disciples before he ascends to the Father. These words offer two interpretative keys for listening to Acts. (1) The Risen Jesus will always be present in the events and preaching in which his discipled witnesses are engaged, albeit sometimes imperceptibly. He remains the principal character throughout Acts and present in what unfolds. As Petersen notes,

> The ascension creates for Jesus the possibility of an altogether different form of presence. The heavenly Christ influences Luke's story at pivotal points, sending the Spirit (2.33), inspiring preachers (7.55-56), calling and commissioning his witnesses (9.5-6; 22.14-15, 17–21), healing the sick (9.34), and using his servants to accomplish his saving purpose in the world (26.16-18). The cumulative effect of such references is that 'the heavenly Christ is a key character within the book of Acts'.[17]

---

[15] *AEC*, p. 73. Joseph Fitzmyer interprets the Acts' ascension scene as the 'guarantee of Christian destiny' in 'The Ascension of Christ and Pentecost,' *TS* 45 (1984), p. 425. It is also the guarantee of Earth's destiny.

[16] See also S. G. Wilson, 'The Ascension: a Critique and an Interpretation,' *ZNW* 59 (1968), pp. 269–81.

[17] David Petersen, *The Acts of the Apostles* (Grand Rapids, MI: Wm B. Eerdmans Publishing Co, 2009), p. 48, with the quote from Matthew Sleeman, 'The Ascension and the Heavenly Ministry of Christ', *The Forgotten Christ: Exploring the Majesty and Mystery of God Incarnate*, ed. Stephen Clark (Nottingham: Apollos, 2007), p. 158.

(2) In the period which follows the disciples' Pentecost experience, for the rest of Luke's story after Acts 2, the Spirit acts as the revealer of God's power and the harbinger of holiness. Most importantly, the Spirit is the agent of fecundity.

Throughout Acts, the Spirit continues to act, giving birth, freshness, and creativity in the disciples' mission, evident in Peter (those who take the message beyond Jerusalem and Judea), and in the words and deeds of Paul. The Spirit's agency allows for a new, unimaginable network of relationships, human and non-human, organic and non-organic, to grow. This ecological perception of Luke's story might be unintended by the author of Acts. However, a fresh intertextual listening to this ancient writing offers new insights for contemporary Jesus followers concerned about our planet.

3

Acts 1.12–2.47. The Ecologically Renewed Household

After Jesus' ascension the disciples go to Jerusalem to await the promise of the Holy Spirit. They return from Mount Olivet, the same topographical feature dominant in the gospel's final chapters. It is from Olivet that Jesus rides humbly into Jerusalem welcomed by the disciples' Earth-Heaven song of peace (Lk. 19.29-44); it is the place to where Jesus returns after his temple teaching (Lk. 21.37-38), and where he prays with his disciples and communes with Earth before his arrest (Lk. 22.39-46).[1] It is not without significance that it is from Olivet that the Risen Jesus ascends to God with his disciples watching on. The attentive listener to the gospel associates ecological meaning with 'Olivet'.

The 'Upper Room' (1.12-26)

The disciples return to the 'upper room' (1.13), the place where they had been staying in Jerusalem. This is a different expression from the identified 'guest room' to which Jesus sends Peter and John to go to prepare for his final Passover meal with his disciples (Lk. 22.11).[2] In the gospel story, Jesus instructs them to enter Jerusalem, follow a man carrying a jar of water to a specific house and enquire from its owner about the location of Jesus' Passover gathering with his disciples. Luke continues, 'He will show you a large furnished upper room.' (Lk. 22.12a).

Luke imagines that the place to where the disciples return after Jesus' ascension in Acts is the same room where they initially gathered with Jesus before his passion. This room, the place about to be overshadowed by God's Spirit, is an ecologically resonant space. It combines the memory of Jesus' eating and its association with food – a consistent theme throughout the gospel with all the Earth related imagery associated with this memory – and the immanent descent of the fructifying Spirit onto the room

---

[1] On the significance of Mt Olivet, see *AEC*, pp. 9, 72, 245, 247.
[2] Reference to the expression 'guest room' occurs twice in the gospel. It appears first in the birth narrative where Jesus does not stay in the 'guest room' because he is with family members (Lk. 2.7). The evangelist later describes a 'large furnished upper room' (Lk. 21.12) with the same expression. Here in Acts, the 'upper room' presumes a larger room, consubstantial with the gospel's gathering place, but large enough to accommodate the envisaged gathering of Jesus' disciples.

and those who now gather in it. In the gospel account, the room is associated with sustenance, communion and failure. As the disciples eat and drink, Jesus foreshadows his betrayal (Lk. 22.21-34). Similarly, in Acts, as the disciples return to the upper room, the evangelist parallels themes of communion and failure. Luke mentions the names of the male disciples, but only eleven of them. This number recalls an incomplete Twelve. It is a reminder of the failure of one of their number, Judas, who no longer accompanies them. The evangelist describes the sense of communion which the eleven share with the women as they pray with, 'one accord with the women, and Mary the mother of Jesus, and with his brothers and sisters (*adelphos*)' (1.14).[3]

The prayer of this reassembled group of discipled leaders occurs in communion with women. A communal and inclusive spirit of prayer also preoccupies Jesus' mother, Mary. This is the first time since the birth narrative that she is explicitly mentioned. Her presence, and the memory of her relationship to Jesus as the fruitful bearer of Earth's Child, is associated with her description as Jesus' mother. Again, here, as in so many of the images that Luke uses in these opening verses of Acts, ecological connections to the gospel are clear for the contemporary auditor.

In this context of prayer and in an inclusively gendered familial space in which Jesus' disciples, including Mary, gather, Peter assumes a position of leadership. The speech which now follows (1.15-22) seeks to bring about the process by which Judas is replaced and the Twelve become re-established. This reconstitution of the Twelve becomes an urgent matter. The number signified the restoration of Israel's Spirit-embraced people. The Twelve need to be restored before Pentecost with its outpouring of the gift of the Spirit that will empower this newly constituted group for mission.[4] Peter's speech highlights and repeats the credentials which Judas' replacement must have. These characteristics concern the eligible candidate's quality of solidarity with the community of disciples and as a witnessing companion to Jesus' whole ministry, from his baptism (Lk. 3.21-22) to his ascension (Lk. 24.50-53; Acts 1.9-11). Matthias is finally appointed and completes the Twelve 'apostles'. Luke's description of the Twelve as 'apostles' further underscores their status as foundational witnesses to Luke's householders of Jesus' ministry, death, resurrection and ascension. The description also affirms their continuing role as witnesses in an Earth-related mission, as seen in the previous chapter.

Pentecost (2.1-36)

The next event that takes place in the 'upper room' where they are together and united, is a household event that affects and influences those gathered. It also establishes the spiritual foundation for what will unfold in the rest of Acts. The event is set on the day

---

[3] The *adelphoi* (the plural of *adelphos*) is literally translated as 'brothers' but it is an inclusive rather than a gender exclusive designation. In the gospel, the evangelist presumes Jesus' larger family. See also I. Howard Marshall, 'Brothers embracing Sisters?,' *The Bible Translator* 55 (2004), pp. 303–10.

[4] Johnson, *Acts*, p. 39; Jacques Dupont, 'Le douxième apôtre (Acts 1.15-26): à propos d'une explication récente,' *The New Testament Age: volume 1*, ed. W. C. Weinrich (Macon, GA: Mercer University Press, 1984), pp. 139–45; Gerhard Lohfink, *Jesus of Nazareth: What He Wanted, Who He Was* (Collegeville, MN: The Liturgical Press, 2012), p. 67.

of Pentecost, the 'Feast of Weeks', one of three important pilgrimage feasts that originates from the nomadic and agricultural period of Israel's life. Of all the feasts, this one reverberates most ecologically.

On this feast, Earth's first fruits, goats and lambs are brought forward, offered to God, in thanksgiving for what the Earth offers (Exod. 23.16; 34.22; Lev. 23.15-21; Deut. 16.9-12). In this explicitly agricultural and Earth-related setting, God acts upon the household of gathered disciples as the Spirit 'fills' them. The visual and audio phenomena described in meteorological terms ('mighty wind', 'fire') that accompany the Spirit's presence are a reminder of the covenant event at Sinai (Exod. 19.16-19) and underscore the power of an experience that orients the future mission of the gathered male and female disciples. This Spirit that formulates covenantal identity at Sinai in Exodus (Exod. 23) 'overshadows' the mother of Earth's Child (Lk. 1.35), is promised by the Risen Jesus (Lk. 24.49), and now empowers Jesus' Earth-oriented agents. The Spirit authorizes them to speak to all Mediterranean diaspora Jews gathered for the feast in Jerusalem.

Peter's speech which follows (2.14-36) seeks to give clarity to his universal audience's experience of God's Spirit, as he draws on the prophet Joel to explain the phenomena. Peter's reshaping of Joel (2.17-21) provides the future mission of the disciples in Acts. This does for the story in Acts what Jesus' interpretation of Isaiah offers to Luke's Christology in the gospel (Lk. 4.18-21). Both are programmatic, explicitly pneumatological, Earth-oriented and implicitly universal; the prophetic quotes concern the whole of creation. Jesus' use of Isaiah can be heard from an ecological perspective. The Earth is part of the liberating mission of Jesus to the 'poor' and to the disciples' concerned about environmental engagement.[5]

'All Flesh ... Shall Prophesy'

In Acts, Luke integrates Joel 3.1-5 (from the Hebrew text) to refer to the 'last days', keeps the syllogistic structure of the first part of the LXX Joel and adds *'and they shall prophesy'* in the stanza's last line (italicized below). For the evangelist, God's voice declares,

> A. I shall pour out my Spirit upon all flesh (*sarx*),
>   And your sons and daughters will prophesy
>   And your youths will see visions
>   And your elders will dream dreams
>   And, yes, upon my male slaves and female slaves
> A¹. I shall pour out my Spirit – *and they shall prophesy.*
>
> 2.17-18[6]

The two phrases, 'I shall pour out my Spirit' (A, A¹), create the frame in which specified actors ('sons', 'daughters', 'youths', 'elders', 'male slaves' and 'female slaves') experience the

---
[5] AEC, pp. 112-14.
[6] Emphasis added.

action of this Spirit that results in certain ecstatic and visionary phenomena ('prophesy', 'visions', 'dreams'). The framing phrases (A, A¹) about the outpouring of the Spirit have additional expressions. The first ('upon all flesh') is original to Joel, the second ('and they shall prophesy'), added by Luke. The second is a logical extension of the first. Together, Luke emphasizes that 'all flesh' 'shall prophesy'. It would seem that Joel's nominated actors ('sons', 'daughters'...) contained by the outer frame are those who represent 'all flesh'. They shall be the ones who prophesy.

The Greek expression for 'flesh', *sarx*, implies an added layer of relationship that is more than anthropocentric. *Sarx* emphasizes humanity in its relationship to the wider environment and world in which human beings live.[7] The expression underscores a person's environmental network of relationships that makes one human. This is testified in the First Testament's Hebrew *basar* (for example, Ps. 63.2). Luke's adoption of Joel's 'all flesh' is not exclusively anthropocentric, it also implies the non-human world.[8] This is clear from Luke's use of the same expression earlier, in the gospel, in interpreting the ministry of John the Baptist by adding to Luke's Markan source, the LXX Isa. 40.5, 'And all flesh shall see the salvation of God.' (Lk. 3.6). The Jewish historian Josephus (37–100 CE) also interprets 'all flesh' in universalistic non-human terms.[9]

If this wider world beyond the human is implied in *sarx*, here in Acts and echoed in the evangelist's use of it in the gospel, then Luke is affirming that the prophetic voice can also be the non-human world, which assists human beings to know themselves in the world in which they live. A similar relationship is not unknown in other Second Testament writings. Paul, for example, attests to the 'groaning' of creation in Romans 8, as it seeks to accompany human beings in their 'groaning' and desire for redemption. Creation participates in God's redemptive act. For Luke, a similar agency is potentially at work in the non-human world (embraced by 'all flesh') which becomes a prophetic voice of God's presence and action through the presence of the Spirit. In this case, the Spirit is not about vivifying human beings exclusively, but every organism that constitutes Earth.

The Joel quote is important. It becomes the interpreting filter for the rest of Peter's long speech, the first of many in Acts that bear the hallmark of Luke's missionary theology.[10] Peter attests to the ministry of Jesus, his death and particularly his exalted status, the fruit of God's actions and witnessed to by the Lukan Peter's interpretation of

---

[7] Johnson, *Acts*, p. 49.

[8] This is evident in Gen. 6.17 ff. In one manuscript tradition, the Western ('D') manuscript changes 'all flesh' to the plural, thus emphasizing the wider non-human world with its stress on universalism. See Conzelmann, *Acts*, pp. 19–20; *NIDNTT* 1 (Exeter: Paternoster Press, 1975), pp. 671–82.

[9] See AEC, p. 101; Alexander Sand, 'σάρξ', *EDNT* 3, pp. 230–3; Eduard Schweizer, 'σάρξ', *TDNT* 7 (Grand Rapids, MI: Wm B. Eerdmans Publishing Co., 1964), pp. 98–151.

[10] The are eleven major speeches in Acts: 2.14-20 (Peter's speech at Pentecost); 3.12-26 (Peter to the Jews at Jerusalem); 7.2-53 (Stephen before the Jewish Council); 10.34-43 (Peter to Cornelius and friends at Caesarea); 13.16-41 (Paul to the Jews among the Gentiles); 17.22-31 (Paul at the Athens Areopagus); 20.18-35 (Paul to the Gentile Jesus followers); 22.1-21 (Paul to the people of Jerusalem); 24.1-21 (Paul to Felix); 26.2-23 (Paul to King Agrippa); 28.25-8 (Paul to his Roman Jewish audience). These speeches are not summaries of historical spoken deliveries but independent literary creations that integrate First Testament and Jewish writings and express Luke's theological agenda. Overall, Acts' speeches are christologically focused, emphasizing God's salvific intent, inviting their audience to respond (with repentance and baptism). See Marion L. Soards, *The Speeches in Acts: their Content, Context, and Concerns* (Louisville, Kentucky: Westminster/John Knox, 1994); Conzelmann, *Acts*, pp. xliii–xliv; Pervo, *Acts*, pp. 38–9.

First Testament characters and texts. Peter's concluding words acknowledge that his audience, representative of the Mediterranean world, will also receive God's 'gift of the Holy Spirit' (2.38c). This promise of the Spirit, 'is to you, and to your children, and to all those who are far off, those whom the Lord our God calls' (2.39).

These words anticipate the direction in which the Spirit will move in its giftedness. While it is promised to those present in Jerusalem, Luke's Peter recognizes that this Spirit will also be gifted to those 'who are far off', a phrase drawn from LXX Isa. 57.19 implying the Gentile world.[11] When Peter next appears in Acts outside of Jerusalem in Acts 9.32f, he will endorse this embrace and formally initiate the Gentiles into a household of Jesus disciples who were originally Jewish and 'whom the Lord our God calls'. That expansion of the Jesus community has its origins here in Pentecost. It is the work of the Holy Spirit and Peter will be present when it occurs.

The Jerusalem Jesus Household Renewed (2.38-47)

Peter's words lead his audience to repentance (*metanoia*: 2.38) and baptism (2.38, 41), a dynamic observed already in the gospel preaching of John the Baptist (Lk. 3.1-20). *Metanoia* is not simply about a conversion to moral rectitude. If the same theological insight from the gospel applies to interpreting its use in Acts, it concerns an attitude of ecological conversion that embraces all Earth's creatures, human and non-human.[12] Baptism, the expression of this ecological conversion, is symbolized through insertion into water, creation's primeval matter. The act of baptismal insertion into water confirms the disciple's link and commitment to creation.

As Luke explicates the wondrous results that flow from Peter's Pentecost sermon, the vast number of newly converted disciples ('three thousand souls'!: 2.41c) undertake their commitment to the Jerusalem Jesus movement in what is Luke's classic summary that concludes these opening two chapters of Acts. Four key aspects characterize the Jesus movement:

> They remained faithful
> (1) to the teaching of the apostles, and
> (2) to the communion (*koinonia*),
> (3) to the breaking of bread, and
> (4) to the prayers.
>
> 2.42

Each characteristic has ecological implications.

1. The apostolic teaching, what they teach in Acts, finds its roots in Luke's gospel as Jesus teaches his disciples, and especially journeys with them, to Jerusalem (Lk. 13.23–19.27), the gospel's denouement. In this journey, Jesus reminds them

---

[11] Johnson, *Acts*, p. 58.
[12] *AEC*, pp. 98–9.

about the importance of Earth's care expressed in terms of awareness and response to the poor, material asceticism and detachment from possessing Earth's goods.[13]

2. The expression 'the communion' seeks to translate the Greek *koinonia*. The usual sense of this term implies the bond of faith in Jesus and God's action through Jesus that cements the relationship of all members of the Jesus household. Taking our cue from the discussion about Jesus' final meal with his disciples in the gospel (Lk. 22.19), membership in the gospel household includes the Earth community.[14] *Koinonia* also implies ecological communion confirmed in the third element of Luke's Jerusalem household characteristics.

3. 'The breaking of bread' refers to the Eucharistic act which typifies the liturgical worshipping act of Luke's householders. The act of 'breaking bread' and the disciples' communion with the broken bread echoes the memory of all the feeding stories in Luke's gospel, the role which food plays in Jesus' communion with creation, human and non-human, and the spirit of inclusivity that characterized Jesus' table ministry. The action of breaking bread now recalled in Acts also signifies the disciples' communion with food that comes from the seed of Earth's most important fruit. But it also symbolizes their communion with bread that is 'broken', that is, with the broken and abused body of Jesus with which they remain in communion even in this post-Easter period of their ecclesial life.[15] The broken nature of this bread that they eat further implies their ongoing communion with all that is 'broken', suffering and struggling humanity and creation. The celebration of Eucharist in the Jerusalem gospel household is therefore an environmentally oriented celebration.

4. 'The prayers', the final characteristic of the Jerusalem Jesus household, would suggest prescribed prayers. The definite article ('the') before 'prayers' would confirm a specific arrangement of particular prayers familiar to Jesus' Jewish adherents. However, a glance again at the gospel would expand upon this. There are several times in the gospel where the evangelist notes Jesus in prayer (Lk. 3.21c; 4.1; 6.12; 9.28; 10.20-22; 22.39-46). A central moment is when Jesus teaches his disciples an ecologically rich prayer in Lk. 11.2b-4c, Luke's version of the 'Our Father'.[16]

This prayer addresses God as 'Father' acknowledged earlier in the gospel as 'Lord of Heaven and Earth' (Lk. 10.21-22). With this awareness of the creator God addressed and called upon, the one praying longs for God's *basileia-ecotopia* to come upon Earth

---

[13] *AEC*, pp. 198–221.

[14] *AEC*, p. 270.

[15] For more on this sense of brokenness and communion with the abused of the Jesus household and eucharistic hospitality, see Michael Trainor, *The Body of Jesus and Sexual Abuse: How the Gospel Passion Narratives inform a Pastoral Response* (New York: Wipf & Stock, 2017), pp. 102–3, 177–8.

[16] *AEC*, pp. 178–80. For a discussion of the economic and political context for this prayer that emanates from the lips of the historical Jesus and is transformed by Luke, see Paul Babie and Michael Trainor, *Neoliberalism and the Biblical Voice: Owning and Consuming* (New York and London: Routledge, 2018), pp. 110–12.

('may your *basileia* come': Lk. 11.2d) to address the wrongs and ills that permeate the disciple's world, to allow enough of Earth's resources to sustain them without resorting to greed or human control over Earth's resources ('Give us each day our daily bread': Lk. 11.3). This leads naturally to a spirit of openness to God's action in the world and upon the household of disciples praying, that brings freedom and openness to others in need ('And forgive us our sins for we ourselves forgive all who are in debt to us': Lk. 11.4b). The praying community also seek to be released from any spirit of greed or diabolical possessiveness that would compromise their relationship to God, others and their Earth-connectedness ('And do not bring us into testing': Lk. 11.4c).

Luke adds to this summary of the life of the Jerusalem Jesus household, now empowered by God's Spirit that has descended upon it. The evangelist explicates the practical implications of its life (2.43-47). Luke emphasizes the relationship which its members had with its apostolic leadership (2.43) in its expression of the Greek ideal of friendship as members 'held everything in common' (2.44).[17] The members of Luke's ideal gospel household expressed their relationship with each other in the way they showed their attitude to and use of their possessions and Earth's gifts:

> [44] All the believers were together and held everything in common [45] and they sold their goods (*ktēma*) and possessions (*uparxis*) and distributed them according to each one's need, [46] daily continuing faithfully in unanimity to attend the temple, and breaking bread in their own homes, they shared food with great joy and generous hearts.
>
> 2.44-46

Through this long sentence in the Greek, Luke seeks to summarize several features of this fledgling Jerusalem Jesus household. Its ecological asceticism and generosity in the use of Earth's goods (*ktēma*) that cemented the spirit of communion among its members is clear in Luke's description (2.44). Their use of Earth's resources expresses in tangible form Jesus' teaching to his disciples in the gospel: to be ascetically free from avarice and generous in nature with their possessions (*uparxis*), the benefit that has come to them from Earth's gifts. Their spirit of largesse comes to its fullest expression in the way they celebrate the Eucharist with all the ecological implications contained in the expression 'breaking bread' (2.46b) explicated above. Their generous distribution of food (2.46c) through table communion – familiar to the auditor from the gospel's Christological portrait of Jesus' table ministry of inclusivity – further supports this spirit. Jesus' common table expressed the way that Earth's fruits become the means of revealing the generosity of a God in love with humanity and in communion with all creation.

A final note of praise to God, the admiration the household receives from 'all people' and its numerical growth (2.47) conclude Luke's ecologically rich summary of the life of the nascent Jerusalem Jesus movement. The evangelist's idyllic description prepares

---

[17] The standard Greek *topos* of the ideal of friendship, that 'friends hold all things in common', was well-known in Luke's day. For example, Plato (*Resp.* 449C), Aristotle (*Eth. Nic.* 1168B; *Pol.* 1263A), Plutarch (*Amat.* 21 (*Mor.* 767E) and Philo (*Abr.* 235) all attest to this utopian ideal.

for what is about to unfold as the household's Spirit-inspired members, especially identified in Peter and John, begin to undertake the work for which they have been empowered.

Conclusion

The first two chapters of Acts lay the foundation for Luke's second volume. They complete the auditors' orientation to key themes that Acts will subsequently reveal: the link which Acts has to the gospel; the reasserted communion between Heaven and Earth, as the sacred and secular cosmological spheres interpenetrate each other; the gospel's ecological teachings that reappear in a fresh mode or new key.

Significantly, there is Luke's summary of the Jerusalem household (2.42). This summary will again be repeated in 4.32-37. Its repetition shows how Luke regards, or would like to regard, the Jerusalem Jesus movement as a manifestation of the well-known Hellenistic ideal of friendship, which acts with ecological responsibility in its treatment of its members. They share Earth's goods and distribute property and possessions to the needy. They ritually symbolize their largesse of spirit through their commensal practice of 'breaking bread' and their joy in sharing food and practical commensurate life. Flaws soon appear in Luke's idyllic Jerusalem gospel household, especially in the way some resist the implications of a shared life and ecologically ascetical lifestyle (5.1-11). Divisions occur along ethnic lines in its commensal practice (6.1-6). These flaws reveal Luke's desire for a generous common life in a household of Jesus disciples far from perfect with its defects apparent.

Finally, the Risen and ascended Jesus continues to be present to his disciples through the Spirit that now permeates the disciples and their household. The Spirit's presence empowers them for the future ministry as their apostolic witness to Jesus continues. Luke identifies the inner circle of Jesus' disciples, the Twelve, as 'apostles'. This becomes the evangelist's preferred expression signifying their primary apostolic status and missionary focus.[18] The next chapter in Acts begins to unravel the expression of the apostles' missionary and healing work. It commences first in Jerusalem and Judea before it moves to Samaria and, finally, to 'Earth's end' with Paul.

---

[18] The designation 'apostle' occurs over thirty times in the Lukan corpus, but only around twenty times in the remaining Second Testament writings.

4

Acts 3.1–6.7. The Fruitfulness of Earth's Children

Peter and John now assume a leadership role amongst the apostles. Luke notes how they maintain their prayerful relationship with the temple, the powerful ecologically resonant prayer space, Earth's navel and meeting place of Heaven and Earth.[1] On their way to the temple's entrance they encounter a beggar, crippled from birth (3.2-3). Instead of offering alms, Peter heals him. The healing occurs in the name of Jesus and the man's response to what has happened is ecstatic. He walks and leaps accompanying the two into the temple (3.7-9). As a crowd gathers amazed at the man's healing and his ability to walk, Peter addresses them and explains that the source of the man's healing is the crucified and Risen Jesus, God's prophet attested to by prophets of the First Testament (3.12-26).

Peter's Address to his Jerusalem Audience (3.12-26)

Luke's purpose in Peter's speech is twofold. It is designed to identify the source of the man's healing. It is God's act through the agency of Jesus. Though ascended to the Father, Jesus' powerful healing presence continues through the apostles. Luke notes that many Jews are attracted to the Jesus movement. This is evident from the summary in 6.7 and, later, the leaders of the Jerusalem Jesus movement note the 'many thousands among the Jews who have believed' (21.20). Nevertheless, the move away from Judaism and the mission towards the Gentiles, which Paul will undertake later, reveal Luke's perplexity that the Jewish people *en masse* did not join the Jesus movement and embrace God's revelation in Jesus.

This introduces a parenthetic note not specifically focused on the ecological focus of this present commentary. But it is important. Peter's speech announces a thematic echoed in other speeches in Acts. It introduces a Lukan anti-Jewish undercurrent that runs throughout Acts. In this speech, Luke considers Peter's Jewish audience responsible for the death of Jesus (3.13). He describes his listeners as 'killers' of Jesus (3.15) who acted in 'ignorance' (3.17) and do not listen to their own prophets (3.22). Luke understands that the Jewish people are the first of the nations invited to receive God's

---

[1] *AEC*, p. 108. For the connection between the Temple and creation, see Jon Levenson, *Creation and the Persistence of Evil* (Princeton: Princeton University Press, 1988), p. 78.

servant, Jesus, and to 'turn from wickedness' (3.26). These themes will be repeated in other speeches in Acts.

Anti-Judaism in Acts (4.1-23)

Unfortunately, this anti-Jewish thematic present in Acts is one of the several Second Testament contributions that led Christians in the third century CE and beyond to the erroneous belief that Christianity superseded Judaism and justified the Christian charge of deicide against Jews. History has shown us the tragedy that has come from this in the pogroms of Jews in Medieval Europe, their execution and attempted extermination under Nazism in the Second World War, and anti-Semitic acts perpetrated against Jews today. One Jewish scholar commenting on Acts remarks,

> Acts makes clear that Scripture, properly understood, foretells who Jesus is and how people will come to respond to him ... Jews respond to these teachings in different ways, some accepting what the apostles preach (2.41; 13.42) and others rejecting it (7.54; 17.2). The book culminates with Paul's fiery denunciation of Jewish unbelief, punctuated by a quote from Isaiah that predicts Jewish intransigence and justifies the opening of God's promises to Gentiles (28.25-28). Luke-Acts depicts the believers in Jesus as possessing the proper understanding of scripture, obedient to God, and serving as the true recipients of the divine promises and blessings. Jesus' followers point to the fulfillment of prophecies (e.g. 2.14-28) as they exhort Jews to accept Jesus as the Messiah. Some Jews, however, generally prove unwilling to convert or are incapable of comprehending God's actions. In contrast the ease with which many Gentiles, including God-fearers, come to this recognition casts further condemnation upon Jews for their unbelief.[2]

This note on Luke's anti-Jewish tendency evident in Peter's speech stands in tension with Luke's ecological themes, derived from a Jewish consciousness, which appear sometimes explicitly and other times subtly. The gospel and now Acts is the fruit of Luke's awareness of the Jewish foundation of the Jesus movement. The evangelical disappointment registered by the author and contained in the speeches will continue. The subtlety of Luke's position on the Jewish question remains enigmatic.

The speech that witnesses to Jesus and the resurrection from the dead is interrupted by the temple authorities (4.1-2) who arrest the two. Luke's brief note of evangelical success seen in the number who come to faith and the popularity of the word (4.4) leads

---

[2] Gary Gilbert, 'The Acts of the Apostles', in *The Jewish Annotated New Testament: New Revised Standard Version*, eds Amy-Jill Levine and Marc Zvi Brettler (Oxford: Oxford University Press, 2011), pp. 198–9. For a further summary of Luke's anti-Jewish bias in Acts, see Warren Carter and Amy-Jill Levine, *The New Testament: Methods and Meanings* (Nashville: Abingdon Press, 2013), pp. 101–7. Tannehill in *Narrative Unity* writes, 'the speeches of Peter in Acts make clear that the end of Luke left another unresolved issue: the rejection of Jesus by the people of Jerusalem and their leaders. This is not to be dismissed as an unimportant matter that can simply be forgotten' (p. 7). See also Robert C. Tannehill, 'Israel in Luke-Acts: a Tragic Story', *JBL* 104 (1985), pp. 69–85, esp. p. 74.

to a further examination from the authorities. Peter's second speech to his interrogators again focuses on the power to heal that comes from the Risen Jesus (4.8-12). The witness of the two apostles, the conviction of the people who have listened to them and the evidence of the presence of the healed man, prevents the Jewish religious leadership from punishing them (4.21-22).

The God of Creation (4.24-35)

When Peter and John returned to their company and reported what had happened, all who heard what the two had to say, prayed to God. Their opening words to God first acknowledge God's creative role in all creation. They address God as 'Sovereign Lord, you who made the Heaven and the Earth and the seas and everything in it' (4.24). This acknowledgement of God's power and presence throughout all creation becomes the primary image of God that the Jerusalem householders applaud. This recognition further ameliorates the threat experienced by the primary apostolic representatives from the religious authorities and 'the kings of Earth' (4.26). It offers consolation to the Jerusalem members who realize that this threat to their faithful witness of Jesus, God's anointed, will be ongoing and lead to persecution, even death (4.28). In their acknowledgement of loyalty to the God of Creation, the Jesus householders seek the virtue of 'boldness' (*parrēsia*) – a theme that appears several times in Acts[3] – as they witness to the gospel, deal with resistance from others, and respond to unforeseen difficulties (4.29).[4] The prayer of the apostolic assembly is answered in an event reminiscent of their earlier Pentecost experience: 'And when they prayed the place in which they were gathered shook, and all, being filled with the Holy Spirit, spoke the Word of God with boldness (*parrēsia*).' (4.31). In the aftermath of the first Pentecost experience, Luke offered a summary of the characteristics of the household of disciples (2.42-47) explored in the previous chapter. After this mini-Pentecost moment, a similar summary occurs (4.32-37).

This second summary explicates the focus of the apostolic witness on Jesus' resurrection. It also reiterates the common life shared by gospel members and expounds further on the implications of their sharing of possessions and resources. Ecological resonances permeate Luke's description. Their primary quality of identity and communion comes from an ascetical attitude towards Earth's goods from which they feel free to be released. The monies that come from this are given to the apostles to distribute to those in need. Acts 4.32-35 is worth quoting in full for a very good reason.

As set out in Figure 5, Luke's reiterated description of the Jerusalem Jesus household reveals a theological depth indicated by its chiasmic-framing structure. The centrepiece of this text is verse 33, concerned with the apostolic witness to the Risen Jesus. This is

---

[3] *Parrēsia* occurs throughout this scene in 4.13, 29, 31. It also occurs in 2.29 and in the final verse of Acts, in 28.31 where, towards the end of Chapter 13, *parrēsia* will be discussed in greater depth.

[4] On the virtue of *parrēsia* and its importance for the Jesus followers in dealing with difficulties within and without the household, see Paul Minear, 'Dear Theo: The Kerygmatic Intention and Claim of the Book of Acts,' *Union Seminary Review* 27 (1973), p. 138.

> ³²The multitude of believers were one in heart and soul, nor did a single one of them call private anything that belonged to them, but they had everything in common.
>
> > ³³And with great power the apostles rendered witness to the resurrection of the Lord Jesus, and great grace was upon all of them.
>
> ³⁴For there was no one in need among them, for those who owned property or houses would sell them and brought the proceeds of the sale ³⁵and placed them before the apostles' feet and it was distributed to each one according to need.
>
> 4.32-35

(A / B / A¹)

Figure 5 Literary structure of Acts 4.32-35.

the heart of Luke's Jerusalem household. The apostles' 'power' and witness in word and deed, the fruit of the Spirit's presence, renders the Risen Jesus present. Because of this presence that encounters all the baptized members of Luke's household, communion in heart and mind comes about (verse 32). Their release from material acquisitiveness addresses the need of the poor (verses 34-35), a primary discipleship quality extolled by Jesus in the gospel (Lk. 9.1-17; 12.13-21).[5]

Generosity and Greed (4.36–5.11)

The next two scenes illustrate how this commitment to generosity manifests itself in two contrasting actions. In Acts 4.36-37, Joseph-Barnabas, a Cypriot Jesus member, sells a family plot of land and lays the money gained from its sale at the apostles' feet. Barnabas' generosity contrasts the deception and greed by two others, Ananias and Sapphira, in the next scene (5.1-11). They, too, sell their property but deliberately hold back part of its sale from the apostles. Their deception and greed result in their deaths. To the contemporary auditor this is a remarkably harsh punishment for deception. But behind the Ananias-Sapphira story lay other theological and biblical motifs which illuminate the judgement on the two who deceive and seek to impugn the holiness of the Jesus household. A key point comes in 5.4 where Peter, who acts like Jesus as a prophetic interpreter, offers the reason for the harsh judgement. The NRSV translates this verse thus:

> ªWhile it remained unsold, did it not remain your own? ᵇAnd after it was sold were not the proceeds at your disposal? ᶜHow is it that you have contrived this deed in your heart? ᵈYou did not lie to us but to God!
>
> 5.4

---

[5] See *AEC*, pp. 151-7; 183-90.

This translation of verse 4a+b seems to suggest that no matter what happened in the sale of the property, the right of Ananias and Sapphira to dispose freely of their property remained, without reproach. The recrimination that comes to them in the final part of the verse (4c) does not seem to fit the preceding part of the verse. Verse 5.4 might be more clearly translated:

> ªSurely, while it [the money you held back from the sale] remained in your possession was it still yours? ᵇAnd having sold it, did it not remain in your control? ᶜHow did you contrive this deed in your heart? ᵈYou have not lied to human beings but to God.

The issue concerns the real ownership of Earth's goods. While property and goods belong to elite members of the Jerusalem Jesus movement, Jesus' gospel teaching (Lk. 12.20) emphasized that such ownership was rather an act of divine entrustment for the distribution to all who are in need.[6] The theological implication of the actions of Ananias and Sapphira is explicit in 5.4c and d. Their action in holding back property or some of the proceeds of its sale is deception. They act as though they are giving over all their goods for apostolic distribution. But their deception reveals their greed. It is an act of the heart which is against God. Their action has moral consequences for which they pay with their lives. Their greed is an act of death. As Tannehill suggests, their death, 'clearly emphasizes the seriousness of the threat of the church represented by Ananias and Sapphira. It is a threat of a particular kind, one that arises from inside, is deceptive, and attacks a central aspect of the church's life, as presented by the narrator: the heartfelt devotion to other demonstrated in the community of goods'.[7] The act of property release, voluntarily encouraged, ensured that unbalanced relationships, typical for ensuring the maintenance of hierarchical social status in the wider Greco-Roman world, were not typical within the household of gospel members who came from different socio-economic situations. They were to have all in 'common' (*koinos*). The wealthier were to make friends with the poorer members of the gospel household. The spirit of communion (*koinonia*), encouraged through the presence of God's Spirit, set the Jesus household apart.

Up until this stage in Luke's story, the household of Jerusalem disciples were of 'one mind and soul' and held all things in common (4.32). There was no material need among them (4.34). Material acquisitiveness and greed were absent. Ecological asceticism and environmental sensitivity symbolized by the voluntary sale of property and goods typified this community. However, the deaths of Ananias and Sapphira, which resulted from the kind of greed that Luke's Jesus indicated as death-dealing (Lk. 12.20-21; 14.2-6), reveal that there are struggles *within* the Jerusalem Jesus household.[8] Things are not as idyllic as the evangelist painted earlier. Perhaps this mirrors the issues facing the Lukan household of the late first century CE.

---

[6] *AEC*, p. 187.
[7] Tannehill, *Narrative Unity*, p. 79.
[8] *AEC*, pp. 141, 160, 167, 183-7, 220.

Witness, Tensions and Growth (5.12–6.7)

Luke contrasts the Ananias-Sapphira incident with what follows. The evangelist notes (5.12-16) the honour in which the Jerusalem Jesus movement is held, the growth in its numbers, its inclusivity ('of men and women': 5.14), and the healing and exorcisms that come from Peter. The spirit of the Risen Jesus continues through the apostles' witnesses after being imprisoned and then miraculously released to preach about 'all the words of this life' (5.20), even to the way they witness to the religious leadership in word and deed (5.17-39). Their witness concerns Jesus' exaltation and the invitation to their audience ('Israel') to repent (5.30-31). Despite being flogged, they leave the presence of their antagonists and continue their preaching (5.40-42). Notwithstanding the humiliation the apostles endure, Luke concludes this episode affirming their faithful adherence to their Christocentric preaching publicly ('in the temple') and privately ('in their own homes'): 'Every day in the temple and their own homes they did not cease teaching and proclaiming Jesus as the Christ.' (5.42).

The growth in numbers in the Jesus household brings with it its own concerns. The story that now follows further hints at how internal tensions continue to preoccupy the fledgling Jesus movement. They tarnish Luke's idyllic portrait (6.1-6). This time the issue concerns the relationship and attitudes between Hebrew and Greek Jesus followers brought about by the increasing number of Jesus followers (6.1a). The problem seems to hinge on language and culture.[9] This is evident in the deferential preference shown to the Hebrew widows and the neglect of the Greek-speaking widows (6.1b) who are being overlooked in what the NRSV translates as 'the daily distribution of food' (6.1c). The Greek literally reads, 'The Hellenists murmured against the Hebrews because in the daily ministry (*diakonia*) their widows were being neglected.' (6.1b, c).

Luke's clue to the real issue here is found in the gospel, in the welcome and hospitality shown to Jesus by Martha and Mary (Lk. 10.38-42). Here in Acts and the gospel, the concern is about the style and nature of *diakonia* at a time when there is great demand and so few engaged in the act of *diakonia*.[10] From the gospel we learn that the main focus of *diakonia*, the public ministry of Jesus' disciples (symbolized in Martha's concern over *diakonia*), is the Word of God. The expression of this focus is in table ministry (6.2c) which translators short-circuit in terms of 'food distribution'.

In Lk. 10.38-42 and Acts 6.1-6 there is an allusion to Eucharistic celebration. This is implied in the meal context in which 'the Lord' comes into the women's house and explicated in Acts by the expression which the Twelve use about 'serving (*diakonein*) tables' (6.2c). If this Eucharistic context is in the mind of Luke in Acts 6, then the neglect which the Greek widows experience is more serious than it first seems. It is a neglect concerning Eucharistic practice. The Greek-speaking poorer members of the Jerusalem community (the 'widows') are being Eucharistically ignored.

---

[9] Tannehill, *Narrative Unity*, p. 81.
[10] Without rehearsing here a comprehensive interpretation of Luke 10.38-42 (see *AEC*, pp. 175-8), I suggest that the main point in the gospel is that *diakonia* is at the service of the Word of God rather than the table ministry.

Jesus' meal-ministry and table communion lay at the heart of his expression of God's *basileia*. They have ecological implications. The gifts of Earth, the fruits of creation generously shared, are symbolically incorporated into all the feeding stories which dot Luke's gospel. They become most significant in Jesus' final meal with his disciples before his death (Lk. 22.14-23) and in the two Easter meals which the Risen Jesus shares with his two disciples at Emmaus as he 'breaks bread' with them (Lk. 24.30-31) and later, with the rest of the gathered but disbelieving disciples, as he eats fish (Lk. 24.42). If the Eucharistic allusion in this story in Acts is accurate, then the neglect of the 'daily *diakonia*' of one of the cultural groups in the Jerusalem Jesus household concerns Eucharistic hospitality. The founding Jewish members of the Jerusalem gospel household seem to exercise a dismissive priority and leadership over the non-Jewish, Gentile members.

The story of Acts 6 exposes a deep underlying issue: purity. This tension evident in this early part of Luke's story of Earth's children will find its careful resolution later in 10.1–11.18 and 15.1-35. There the apostles under Peter's leadership will determine how to resolve the difficulties of table communion between Jewish and Gentile Jesus followers, especially in the celebration of the Eucharist, as they discern ways of preserving purity regulations and kosher food laws. This later resolution will lay the foundation for a robust concentrated mission amongst the Gentiles taken up by Paul in the final part of Acts. This progression of the gospel message beyond Judaism begins to be anticipated from now until Paul's commission with the growth and development that flows from Acts 6 as the gospel is preached in Samaria and Philip's baptism of the Ethiopian eunuch (Acts 8).

In Acts 6, the Twelve demonstrate a freedom in ministerial flexibility that addresses the present situation. Pastoral need determines specific ministerial expression, rather than a pre-defined ministerial *diakonia* determining how to address the pastoral issues. The apostles formally appoint seven to 'wait on tables' so that they can dedicate themselves to 'the prayer and the ministry (*diakonia*) of the word' (6.4). Those appointed have Greek names, suggesting that their focus will be to the Greek-speaking members of the Jesus household. The seven's diaconal service to 'wait on tables' helps free the apostles to the *diakonia* of the Word. Thus, ministry (*diakonia*) has two expressions: one to the service of the Table, the other of the Word. The two together represent the narrator's household experience of worship, of Word and Table, reflective of the Eucharistic practice in Luke's day. This reflection in Acts, of the two elements of Table and Word, is an echo of the event in Luke's gospel journey narrative in which Jesus enters into the house of Martha and Mary in Lk. 10.38-42. A similar memory is seen in the feeding story of the hungry crowd in Lk. 9.10-17, and captured in the word-table dynamic present in Luke's Emmaus story (Lk. 24.13-35).[11]

After the issue of pastoral neglect is addressed with the appointment of the seven, Luke notes again the growth of God's word reflected in the number of followers and the reminder that even Jewish priests join the movement (6.7): 'And the Word of God grew (*auxanō*), and the number of disciples in Jerusalem increased (*plēthunō*) greatly,

---

[11] *AEC*, pp. 285–9.

and a great crowd of priests submitted to the faith.' (6.7). Despite frustrations that beset the fledgling Jerusalem Jesus household, growth continues. Generosity, rather than greed, sharing of Earth's resources, rather than the exclusion of certain ethnic groups, growth, rather than decline, attraction, rather than repulsion and rejection: all these characterize Luke's optimistic portrait of this first and second generation of Jesus followers. The writer has established the pattern that will continue through the rest of Acts: regardless of everything that seeks to conspire against the expansion of the Jesus movement beyond Jerusalem, God's word will be ultimately fruitful to Earth's end. Generativity rather than sterility will be Luke's catchphrase.

Conclusion

The expressions for the growth of the Jerusalem Jesus movement in terms of 'growth' (*auxanō*) and 'increase' (*plēthunō*) that conclude this section of Acts reveal the ongoing influence of Luke's meta-parable of the sowed seed in Lk. 8.4-21. This parable anticipates and defines the reaction to Jesus' ministry and the eventual fruitfulness of what emerges from his proclamation of God's *basileia-ecotopia*. The seed, initially frustrated by the environment into which it falls, finally reflects the superabundant fertility of God's presence. That same pattern is repeated in the story of Earth's children.

The abundant harvest of the word through the preaching of the community of the disciples is evident in the large numbers of converts baptized on the day of Pentecost (2.37-47). The fruitfulness of the disciples' preaching is also evident through the summaries that mark the early chapters of Acts (4.32-37; 5.12-16; 5.42; 6.7; 8.1b-8). All this encourages optimism in the audience addressed by Luke and establishes the future theological agenda which will unfold: God is faithful and can be trusted; God's word will be ultimately fruitful and eternally unstoppable; the household of Jesus followers will flourish despite civic opposition, religious persecution, even martyrdom that is about to be experienced by one of the seven.[12] Nothing will frustrate the plans of God as the Jesus movement continues to grow and flourish. Ecological imagery from the parable of Earth's Child concerning the sowed seed summarizes the fruitfulness of Earth's children in Luke's second volume.

---

[12] J. Njoroge Wa Ngugi, 'Stephen's Speech as Catechetical Discourse,' *Living Light* 33 (1997), pp. 64–71.

5

Acts 6.8–8.1a. Earth's Presence in Stephen's Story of Israel

Two of the seven, Stephen and Philip, appointed by the apostles in the immediately preceding scene (6.1-7), now play a significant part in the narrative plot in Acts. Acts 6.1–8.3 brings us to an important strategically located section of the book of Acts.[1] Despite the apparent frustration experienced by members of the Jesus movement, and now, the imminent martyrdom of Stephen, gospel fidelity and witness will reap its own rewards. This section also introduces us to Saul (later, called 'Paul') who will occupy most of the remaining chapters of Acts.[2] His presence at the execution of Stephen towards this section's conclusion sets up his later conversion to a religious movement that he first persecutes. It prepares for his subsequent missionary endeavours. Unlike Paul who dominates the second half of Acts, Luke only allows Stephen's story two chapters. But his function for Luke's theology is inestimable. Stephen is presented as a wisdom figure of grace and power, a revealer of God's word, and a worker of 'signs and wonders' (6.8).

Stephen, the 'Wonder Worker' (6.8)

The wisdom expressed through Stephen functions at a deeper level in the narrative. Stephen's wisdom is an expression of the divine wisdom available to Luke's gospel household and revealed earlier through the ministry of Jesus. It is a wisdom unable to be silenced or defeated by Diaspora Jewish officials. It is powerful and effective. This wisdom motif provides Luke with one parallel between the career and ministry of Jesus in the gospel and Stephen's brief appearance in Acts. A second parallel is found in the death scenes of Jesus and Stephen. The language in which Luke casts Stephen's final act of forgiveness and his dedication to God at the moment of death clearly echoes the crucified Jesus.[3]

---

[1] Delbert L. Weins, 'Luke on Pluralism: Flex with History,' *Direction* 23 (1994), pp. 44–53.
[2] The author of Acts almost surreptitiously begins to call Saul 'Paul' from 13.9. This remains Saul's changed name for the remainder of Acts. For further comment on the name 'Saul', see Michael Kochenash, 'Better Call Paul "Saul": Literary Models and a Lukan Innovation,' *JBL* 138 (2019), pp. 433–9.
[3] Ben Witherington, *The Acts of the Apostles: a Socio-Rhetorical Commentary* (Grand Rapids and Cambridge: Wm. B. Eerdmans, 1998), p. 253.

Throughout the whole Stephen-narrative, reverberations and thematic parallels with the story of Jesus continue and find expression. Stephen's trial speech also becomes a summary of Lukan theology and orthopraxis. It reveals to Luke's Greco-Roman Jesus followers a way of living in the present and into the future. The story of Jesus is not something lost to the past. It finds continual expression in the lives and deeds of those whom Luke's audience would still remember. Through this narrative dynamic and thematic interplay between the stories of Jesus and Stephen, the evangelist encourages the gospel audience to continue to live out the life and ministry of Jesus in their own lives.[4] For them the story of Jesus must become a rich source of reflection that influences their present. According to Luke, they must live like Stephen in Acts 7, imbued with the spirit and compassion of Jesus, committed to the cause of God and confident in God's support, especially at the most powerful of moments: martyred death.

The Structure and Purpose of Acts 6.8–8.1a

From a literary perspective, Stephen's trial and speech (6.8–8.1a) is framed between two summary statements (6.7; 8.1b-8).[5] The first (6.7) was looked at briefly in the previous chapter. Like the summary that concludes the present section (8.1b-8), this emphasizes the growth that occurs in the Jerusalem Jesus household amid persecution. In this framing technique which Luke employs often, these summaries prepare for and speak to the story of Stephen, which in turn illuminates the reason for the suffering that accompanies the mission of the Jesus movement.

The Stephen story itself (6.8–8.1a) has a literary balance (Figure 6). Two summaries (6.7; 8.1b-8) surround two scenes (6.8–7.1; 7:54–8.1a) that highlight the central feature of the narrative, Stephen's speech (7.2-53). Strong echoes of Jesus' trial and execution from the gospel feature. The first scene focuses on the public religious trial that stems from Stephen's ministry of grace and power (7.8). The second concludes the Stephen

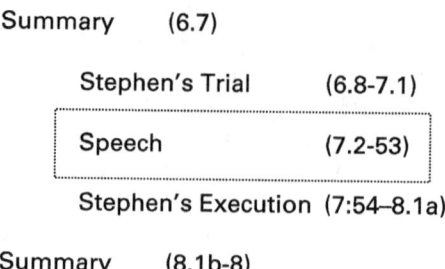

Figure 6 The literary structure of the Stephen story in Acts 6.7–8.8.

---

[4] David S. Dockery, 'Acts 6–12: the Christian Mission Beyond Jerusalem,' *Review and Expositor* 87 (1990), pp. 423–37. I see a clear christological agenda influencing Acts 7 and Stephen's speech, contra John J. Kilgallen, 'The Function of Stephen's Speech (Acts 7:2-53),' *Biblica* 70 (1989), pp. 173–93.

[5] For an alternative structure, see Simon Légasse, *Stephanos: histoire et discours d'Étienne dans les Actes des Apôtres* (Lectio Divina 147. Paris: Cerf, 1992).

story with his public execution and becomes the opportunity for Luke to introduce the principal actor for the remaining chapters of Acts: Paul.[6]

The Stephen story is important for Luke. It narrates the writer's theologically constructed speech and Stephen's martyrdom. The rejection of Stephen which ends in his martyrdom flags what will soon happen to Jesus followers in Jerusalem. The speech which Luke creates for Stephen is the longest in the Book of Acts.[7] This is Luke's reinterpretation of Israelite history. It helps the writer establish the reason for the progress of the Jesus movement beyond Judaism into the Gentile world and towards the West.[8] Stephen becomes Luke's apologetic mouthpiece for responding to the blasphemous charge laid against the gospel audience: it offers a particular viewpoint on Moses, the Torah and the essence of Jewish life.

The speech reveals another agenda besides the writer's perspective on Israelite history. Luke addresses a pastoral issue that has emerged as the household's membership becomes increasingly more Gentile: how faithful can it be to the Jewish origins of its original audience? Can it legitimately move away from Judaism and Jewish cultural practices as it attracts more Gentiles and God-fearers? Are members of the Jesus movement faithful to God's plan or are they renegades?[9] The author seeks to legitimize this new direction and address these issues, unfortunately, as previously noted, with negative appraisal of Judaism. This will become the reason that Luke uses later in Acts when Paul is instructed to focus his mission on the Gentiles, anticipated in the next chapter in Paul's encounter with the risen Jesus (9.1-9). Here his change of heart causes him to seek out Jerusalem's Jesus followers whom he hitherto had persecuted. Luke's need to affirm the theological legitimacy of the embrace of the Gentiles also happens through Stephen's speech. But unfortunately, it will occur at the expense of questioning the validity of the Jewish people in their ongoing covenantal relationship to God.

Returning to the content of Stephen's speech, Stephen does not seek an acquittal of the charges of blasphemy. Rather, Luke portrays him using the accusations as a way of indicting his accusers.[10] The speech addressed to the Jewish leadership offers a Deuteronominstic view of history that reveals the positive understanding of Israel's past and a negative judgement on its present leaders. As Johnson points out, Luke's narration of Israel's past infidelity, similar to interpretations found at Qumran in the *Damascus Rule*, 'helps to legitimate the claims of the present community to be the authentic realization of Israel in the present'.[11]

Through Stephen, Luke's principle concern is to downplay the importance and function of the Temple and underscore the purpose of Jesus' mission.[12] This

---

[6] For an alternative presentation on the literary structure of Acts 7, see Delbert Weins, *Stephen's Sermon and the Structure of Luke-Acts* (N. Richland Hills, TX: Bibal, 1996).
[7] Stephen's speech occupies 365 verses out of about 1,000 verses in total for all speeches in Acts: Soards, *Speeches*, p. 1.
[8] Witherington, *Acts*, p. 252; Peter Dschulnigg, 'Die Rede des Stephanus im Rahmen des Berichtes über sein Martyrium (Apg 6,8–8,3)', *Judaica* 44 (1988), pp. 195–213.
[9] Johnson, *Acts*, p. 119; Kilgallen, 'Function', pp. 173–93.
[10] Tannehill, *Narrative Unity*, pp. 84–5.
[11] Johnson, *Acts*, p. 135.
[12] For a similar judgement that considers Luke's assessment of Judaism as positive, see David Ravens, *Luke and the Restoration of Israel* (JSNT Supplement Series 119. Sheffield: Sheffield Academic Press, 1995). Regarding Luke's attitude to the temple, see Duschulnigg, 'Die Rede', pp. 195–213; Francis D. Weinder, 'Luke, Stephen, and the Temple in Luke-Acts', *BTB* 17 (1987), pp. 88–90.

Christological motive is revealed in the last verses of Stephen's overview of Israel's history: Israel silenced and persecuted its prophets who, 'foretold the coming of the Righteous One, and now you have become his betrayers and murderers' (7.44). In Stephen's interpretation of Israel's history that leads to this final invective against its leaders, Luke uses 'Earth' language. Its location, especially at 7.3-7, 33, 49–50, is significant in the overall development of Stephen's selective presentation of Israel's story. Here appears the ecological importance of the speech and the value which Luke places on Earth.

'Earth' (*gē*) Language in Acts 7

'Earth' (*gē*) occurs in verses 3, 4, 6, 29, 33, 36, 40 and 49 of chapter 7. Its repetition indicates its importance for revealing Luke's 'ecological' sensitivity already identified in Lk. 2.14. There are three main sections of Stephen's speech where *gē* appears: (A) The introductory part of the speech (7.3-7); (B) the directive which God gives Moses as he approaches the burning bush (7.33); and (C) Luke's adaptation of Isa. 66.1 towards the end of the speech (7.49-50). However, it may appear that Earth is not the principal focus of Stephen's oration. Its central message concerns Israel's history, a story of acceptance, care and rejection. It is a parenetic reflection on God's promise, care for the Jewish people, and the importance of Israel's prophetic tradition and its rejection. In this story Earth seems to play a role, an apparent minor one, but its connection to the story of Israel is unmistakeable.

Stephen's speech is in three parts. The first (7.2-19) begins with the story of Abraham, Joseph and the other patriarchs. The second and major part (7.20-34) focuses on Moses, the great leader and prophet of the Jewish people, and highlights God's concern to liberate the people of Israel and consolidate their salvation. This positive portrait of God's redemptive plan is contrasted to the leaders' rejection of God, which occupies the third part of the speech (7.35-53). Their continued resistance to this plan is reflected in their criticism of Moses (7.35-43) and persecution of all the prophets (7.52).

Through this speech, Luke's Stephen lays the ground for the way Jesus and those who follow him will also be rejected. In essence, Stephen articulates Luke's reason for the ultimate failure of the Jewish mission. Given the writer's overall purpose and perspective, Earth plays an important role in Luke's nuanced presentation of Israel's history. It surfaces in the opening verses of the speech that now deserve closer examination (7.3-7). Most of Luke's use of *gē* is found in these initial verses.

Acts 7.3-7

The speech begins with Stephen's account, drawing from Gen. 12.1, of God's address to Abraham:[13]

---

[13] Acts 7.3-6, though, is Luke's reflection on the LXX, especially on Gen. 11.31-2; 12.1-4; 15.13f; 16.1; 48.4; Deut. 2.5 and Exod. 2.22.

³ 'Go out from your land (*gē*) and from your kin and go into the land (*gē*) which I shall show you.' ⁴ Then going from the land (*gē*) of the Chaldeans he [Abraham] lived in Haran. And from there after his father died God removed him into this land (*gē*) in which you now live. ⁵ And he was not granted an inheritance in it, nor so much as a foot's pace, and God promised to give it to him as a possession for him and his posterity after him, though he had no child. ⁶ God spoke in this way, that his descendants would be aliens in a land (*gē*) belonging to foreigners who would enslave and oppress them for four hundred years. ⁷ 'And the nation which they serve I myself will judge,' says God, 'and after these things they shall come out and worship me in this place.'

7.3-7

These verses are packed with either explicit 'Earth' language or inferences to it. These reveal Luke's recognition of the role and function which *gē* plays in the origins of Israel's story of the Patriarch.

Gē, like Abraham and his entourage, needs divine attention. The poetic parallelism of the first two verses belies the objective regard which Luke has for *gē* in the Abraham narrative.¹⁴ In verse 3, *gē* is dissociated from Abraham's family. The Greek clearly brings out the separation between the two entities ('Go out from your land *and* from your kin'). In other words, *gē*, like Abraham's kinship group, is connected to Abraham. There is a symbiosis between Israel's ancestor and the earthly location of his original settlement. The command to 'go out from your *gē*' is an invitation to release himself of proprietorship of the *gē* described as a possession. This reflects the gospel encouragement that Jesus gives his disciples not to possess Earth's goods, but be ascetically free from material possessions and live with a spirit of largesse. Something similar echoes in God's commission to Abraham.

This respect and appreciation for *gē* surface more clearly in the next few verses. The second half of verse 3 ('... go into the land/*gē* which I shall show you ...') affirms that *gē* is the subject of divine action to which God directs Abraham. This new moment in Abraham's story confirms that *gē* is not his possession. He comes to it under God's supervision and its relationship or identity is circumscribed by divine action. Its existence is shaped by God to partner with Abraham, the great patriarch of the Jewish people and, by inference, with all human beings. *Gē* therefore, under God's direction, becomes an indispensable collaborator in human growth and community. Thus, God decrees that *gē* will now be a participant in this new moment, as Israel's story of salvation unfolds through Abraham and his entourage who are directed to a new location on Earth's surface.

---

14 The deliberate poetic parallelism between Acts 7.3 and 7.4 reflects a theological insight which Luke seeks to explore in the figure of Abraham. In v. 3, Abraham is instructed by God to go from his place of birth into a land which will be pointed out to him. There are two main actions commanded by God and both actions are focused on *gē*. Verse 4 shows how Abraham faithfully executes God's command. As God instructs Abraham in two actions in v. 3, so in v. 4, Abraham responds with two actions: he leaves the security afforded by his original place of settlement, and finally arrives under God's direction to the place in which the present audience of Stephen's speech live. The theological import of the sentence could not be clearer: Abraham is God's envoy and faithful respondent, modelling the kind of response and openness indicative of Luke's gospel household.

In these opening verses, *gē* becomes more than an objective landscape on which the divine drama is set. The verses identify Abraham's movement going from one place to another, until finally settling into what will eventually become Israel and the backdrop for the rest of Luke's specific narration of God's promises. Geography is the vehicle for the narrative as Luke specifies the location of the various moments in Abraham's migration. The geographical descriptor, *gē*, is an ecological identifier. The locations for Abraham's sojourn and eventual habitation are not merely GPS coordinates. *Gē* is an environment of living and non-organic beings that assist a human being to connect to the network of life forms and creative matter. Stephen (or God, or Luke) reminds his listeners that this *gē* is the place and environment 'in which you now live'. No matter what events take place that shape the people whom God accompanies in their history, no matter their geography, the *gē* will continue to shape them and influence their lives. Earth-*gē* placement continues to affect God's people.

With verse 4, Earth becomes a boundary that marks off one cultural group or nation from another.[15] It is a geographical locale where Abraham and his family live. This perspective of *gē* as a place of human habitation and ecological communion is reflected in other parts of Stephen's speech (7.29, 36, 40), and intimately connected with the well-known locations associated with Israel's story of salvation: Midian and Egypt. In Midian, Moses encounters God, and Egypt provides the important geographical focus for Moses' liberating leadership.

In verse 4, God's initiative and plan is at the forefront of the writer's narrative and theological agenda: 'Then going from the land (*gē*) of the Chaldeans he [Abraham] lived in Haran. And from there after his father died God removed him into this land (*gē*) in which you now live.' (7.4). *Gē* is again the subject of divine action. God acts to move Abraham from one place to another. God 'removes' Abraham from 'the land of the Chaldeans' to settle in the 'land in which *you now live*'. Here, Luke's Stephen links with his present audience. They are identified with a specific location. They are associated with and can be identified with *gē*. This identity assumes literary and thematic importance.

As the next verses seem to suggest, *gē* suffers in the same way as the implied audience addressed by Stephen. Like Abraham and his kin, God can act upon *gē* to bring about its liberation. It is affected by the quality of the One who acts upon it. It is theologically and anthropologically affective. It can be infused with holiness through God's action, and it can suffer in its communion with the movement of peoples represented through Abraham's migration. Earth's relationship to the human drama means that it is also active in the events that unfold. *Gē* is part of Luke's salvation-oriented geography. It is the narrative context revealing God's salvific intent for Abraham and his family. It represents the geographical and ecological context of occupation from which people leave or to which God directs them (7.3, 4). Theologically, Earth, like God's people to whom it is intimately linked, is in need of redemption to be an agent of release and divine communion.[16] Verse 5 reads, 'and [Abraham] was not granted an inheritance in

---

[15] Hermann Sasse, 'γῆ, ἐπίγειος,' *TDNT* 1 (Grand Rapids, MI: Wm B. Eerdmans Publishing Co., 1964), pp. 677–81.

[16] In those other texts in Acts 7 where Earth appears, but is not explicitly dealt with, Earth needs a miracle (7.36); from it God's people seek to be liberated (7.40).

it [*gē*], nor so much as a foot's pace, and God promised to give it to him as a possession for him and his posterity after him, though he had no child' (7.5). This affirms that the *gē* belongs to God not Abraham. It is not a human possession, not even a 'foot's pace' of Earth. It is not controlled by human beings. However, God promises *gē* to Abraham as a 'possession' for him and his posterity.

Gē: Gift and Reverenced Sacred Agent (7.7)

There are two aspects to this promise that militate against seeing *gē* as a passive object of possession that can be used or commandeered at Abraham's whim. The first comes from the original context for this verse. Luke draws on the Jacob-Joseph narrative in Genesis. In Genesis 48, the dying patriarch, Jacob, speaks to his son, Joseph. He affirms God's promise of land and posterity. According to Jacob, God says to him: 'I will make of you a company of peoples, and will give this land to your offspring after you for a perpetual holding' (Gen. 48.4. NRSV).

Luke stitches the Genesis promise to Jacob concerning land into Stephen's speech. The context of the promise ('for a perpetual holding') assumes that Earth is God's gift to Jacob, not for selfish gain but as a 'perpetual holding'. It is in Jacob's care for future generations. A second insight to 7.5 comes from Luke's teaching about possessions in the gospel already reflected upon. Jesus' instructions to his disciples are clear: they are to release themselves from being possessed by possessions (Lk. 6.30; 12.13-21).[17] Rather, a spirit of largesse must characterize their discipleship. This means that what they have is to be used for the good of others, especially those who are in need (Lk. 8.3). The chapters of Acts leading up to the Stephen narrative demonstrate how this gospel teaching is lived out amongst the Jerusalem household of Jesus followers. What they own is given over to the apostles who use it to relieve the sufferings of the poor (2.44-45; 4.32-37).

These reflections on possessions lie behind Stephen's speech in verse 5 and carry over into verse 6, which demonstrates the opposite to the gospel teaching on possessions. *Gē*, as we know from our experience and the ecological damage caused to creation, becomes a possession co-opted by foreigners. They ecologically abuse Earth to participate in the enslavement and oppression of God's people. In verse 6, foreigners manipulate and co-opt *gē* to enslave and oppress God's people: 'God spoke in this way, that his descendants would be aliens in a land (*gē*) belonging to foreigners who would enslave and oppress them for four hundred years' (7.6). Because of its association with 'foreigners', people not belonging to God, *gē* can be indirectly the cause of tyranny and ill treatment. Through its association with those who do not actually belong to this place, God's people, already alienated from the land, are now persecuted. These attributes reflect the same oppressive characteristics of Jewish leadership that Stephen's speech is intended to address.

Thus, Earth is not a passive object unaffected by what happens within the human community. It is interrelated with those who do good and evil. This affects Earth. For this reason, Earth, like the religious leadership that occupies it, can be abused by humans to become an indirect cause for the crushing disaster experienced by God's

---

[17] See *AEC*, pp. 183-7.

people. The alienation and hardship which *gē* brings about in God's people is further realized by comparing Earth's description in verse 6 ('a land belonging to foreigners') with the source from which Luke drew, LXX Gen. 15.13 ('a land not their own'). In other words, the writer of Acts recognizes the contribution which Earth can make to the human community. It is not neutral.[18]

Acts 7.7 offers a redeeming and balancing picture associated with the place in which Abraham's descendants live. *Gē* imagery is explicit. Through God's agency Earth becomes a place of divine encounter and worship. This liturgical association, a reminder of the angelic hymn of Luke 2.14 in which a similar correlation is made, continues to uphold the initiative of God who determines the value of all created things. This includes Earth.

Verse 7 imperceptibly sums up the way Earth is perceived in this opening section of Stephen's speech: Earth belongs to God and is gifted to humanity who can engage it for good or evil purposes. It has an agency that affects human beings. Its quality is determined by the worth of those who relate, come to or dominate it: whether God, Abraham or foreigners. What seems clear for the writer of Acts 7 is that Earth can be an agent of location for human beings (7.3-6, and represented in the figure of Abraham), and a revealer of God's presence and holiness (7.7). What is implicit in 7.7, that Earth can be sanctified by God, becomes explicit in our remaining two focus texts, 7.33 and 7.49-50.

Acts 7.33

As Stephen continues to trace the story of the patriarchs after Abraham, he moves to consider Moses, whose story now occupies much of the speech. After Luke reshapes the story of Moses from LXX Exod. 2 and 3, and focuses on Moses' encounter with God in a burning bush, God directs Moses, 'Remove the shoes from your feet, for the place upon which you stand is holy earth (*gē*)' (7.33). This is a defining moment in Moses' experience of God and his commission to liberate the enslaved Israelites. In the context of Moses' commission and his sacred encounter, Earth is described as 'holy.' This is the only time in the whole Lukan corpus that *gē* is explicitly invested with sacredness and described in language normally reserved for God. Here Luke expands on the vision of Earth implied in the previous verse, evidenced earlier in the birth narrative and the angelic hymn of Lk. 2.13. In Acts 7.33, Earth is considered so intimately connected to God that it reflects the sanctity of God's very being. For this reason, Moses is asked to remove his shoes. This request adds a further dimension to reveal Earth's sanctity and the reverential stance which Earth's children must have towards it.

In Exod. 3.33, and in Luke's reinterpretation of the event here, the same meaning is given. Moses' body, the consequence of God's creative act in Genesis, must touch the very substance from which it was made. This is Earth (*adamah*, Hebrew) from which

---

[18] For a discussion of Luke's use of the LXX, see Huub Van de Sandt, 'Why is Amos 5.25-27 quoted in Acts 7, 42f.?', *ZNW* 82 (1991), pp. 67-87.

God creates *Adam*, the earth creature. The act of removing shoes is an act of reverence. Earth becomes the location in which human beings, represented in Moses, offer obedience to God. It is the place where humanity's relationship to God, disrupted by the Earth-ling of Genesis, is fully restored.

Earth also reminds the addressees of Acts of the primordial vision of human freedom and communion of Genesis. Verse 7 presents *gē* as offering the possibility of communion with God and reminding Luke's audience of the loving creative act of God that first fashioned human beings. As Stephen presents the central memorable features of the story of salvation, Moses can now be God's agent of liberation. He has encountered the liberating holiness of God experienced in the Earth upon which he has walked. This positive and religious regard in which *gē* is held in 7.7 is further explored in 7.49-50. This comes towards the climactic end of Stephen's speech. In this final section the reason that Earth can be considered holy more clearly emerges. This comes from an inherited cosmological duality that permeates the First Testament and the Mediterranean world of Luke's audience, transformed by Earth's Child and the immersion of Earth with Heaven and the celebration of Heaven on Earth. This realignment has happened through Jesus' ministry, celebrated in the two hymns that frame the gospel (Lk. 2.14; 19.38) and echoed in the ascension scenes that conclude the gospel (Lk. 24.50-53) and begin Acts (1.9-11).

Acts 7.49-50

The final section of Stephen's speech acts as an *apologia* that interprets history from Moses to Solomon (7.44-53) and accentuates the part Moses plays as prophet and a type of Jesus. Moses is presented as one like Jesus, exiled yet empowered through the Spirit to lead God's people.[19] Stephen also reflects on the place of the Temple to argue for God's transcendence.[20] Stephen's reflection offers another perspective of Earth consistent with earlier observations: Earth is the subject of divine action and dependent on God. This estimation also emerges from a cosmological dualism that interprets Earth in relation to Heaven.[21] In the gospel and Acts, this separation of these two cosmological spheres becomes reinterpreted. Heaven and Earth penetrate each other.

Stephen recognizes that Solomon built a house for God (7.47) but argues that God does not dwell in houses made by human hands (7.48). Luke's critique of a Temple-centred theology is also a judgement on a form of human conduct that seeks to control God's action. The temple is the narrative symbol or metaphor for human activity. The temple structure is of human origin and intent but, argues Luke through Stephen, God is not dependent on such a structure. The argument which Luke marshals at this juncture through quoting LXX Isa. 66.1 seeks to reinforce this point of view.

---

[19] Johnson, *Acts*, pp. 136-7.
[20] Dennis D. Sylva, 'The Meaning and Function of Acts 7:46-50,' *JBL* 106 (1987), pp. 261-75.
[21] For another angle of how dualism is interpreted in the New Testament see Thomas E. Schmidt, 'The Penetration of Barriers and the Revelation of Christ in the Gospels,' *NovT* 34 (1992), pp. 229-46, who discusses Stephen's encounter with God as an example of 'divine penetration': the broaching of the spheres separating Heaven from Earth.

> ⁴⁹'Heaven is my throne,
> The earth my footstool,
> What house will you build for me, says the Lord
> Or what place for my rest?
> ⁵⁰ Surely my hand has made all these things?'
>
> 7.49-50

Luke's reworking of the Septuagint original adds 'says the Lord' into the two questions and 'Surely' into the final line of the quotation to render interrogative what was originally declarative.[22] In verse 50 Luke inverts Isaiah's 'For all these things my hand has made …' to 'my hand has made all these things'. This emphasizes God ('my hand') rather than human beings as the agent and initiator of creative activity. Luke's redactions to Isaiah serve to reinforce Stephen's argument. They emphasize divine judgement on the human endeavour to create a place for God and to locate and confine the divine presence.

Luke uses a conventionally inherited understanding of Earth from Isaiah that can be illuminated through an application of the ecojustice principle of intrinsic worth. In Luke's view, Earth is created by God. God determines its existence. It is subject to God's power and judgement.[23] Together with Heaven, Earth constitutes the total cosmos, dependent on God's will. They represent two aspects of creation and divine presence. In an original cosmic duality, Heaven is considered hierarchically superior. It is the place of God's throne. Earth is inferior, as God's footstool and the place where God's feet rest.[24] However, Luke's story of Earth's Child and the redefined convergence of Heaven and Earth are expressions of the divine presence: one ('throne') the seat of divine creative cosmic origins; the other ('footstool') the location of God's powerful presence.

This affirmation of the communion between Heaven and Earth carries over into the final scene of this important section of Acts as it concludes with Stephen's martyrdom. His death occurs as he sees the Heavens opened and the exalted Jesus with God (7.56). Here, one of Earth's children returns to Earth's ascended Child dwelling in God's abode. Stephen's communion with God beyond death is assured. He dies in a manner that reflects the death of Luke's Jesus, in communion with God, entrusting his spirit to Jesus and uttering a final word of forgiveness for his executioners (7.59-60).

Finally, Luke's introduction of the one who will occupy most of the remaining story of Acts, Saul, completes the scene. As Stephen dies entrusting himself to Earth's Child, those who stone him, 'laid their garments at the feet of a young man named Saul … and Saul approved of their killing him' (7.58; 8.1). The last time the attentive listener to Luke's writings would have heard something like this action of placing garments before another was on Jesus' entrance into Jerusalem (Lk. 19.36). There, the garments belonging to the disciples are thrown over the colt upon which Jesus rides into Jerusalem (Lk. 19.35) and onto the path along which the Jesus-bearing colt moves

---

[22] Johnson, *Acts*, p. 133.
[23] Sasse, 'γῆ, ἐπίγειος,' p. 679.
[24] Habel, *Readings*, pp. 123-36.

(Lk. 19.36). At a deeper level of narrative symbolism, the path and the colt are covered with garments which gave identity to the disciples. They, Earth's road and creature, become discipled and symbolically accompany Jesus to a death that will bring about liberation for all God's people and creation.[25]

Here, in Acts, in a reversed image, the clothing that identifies Stephen's killers, a creation or gift of Earth, is placed before Luke's main character who initially will bring about the persecution of Earth's children.[26] Clothing is the product of wool or flax (in the case of linen) and accompanies Luke's characters in Acts. Clothing clothes; it identifies, honours or, in the case in the present scene, is co-opted in and witnesses to Stephen's execution. Earth indirectly becomes an actor in the event, albeit unnoticed and implied.

Many of the characters of Acts are full, respected and engaging (for example, Peter and John in 3.1-26; Paul from 9.1 onwards; Stephen in 6.8-7.58; Philip in 8.26-40); others are more shady, and antagonistic towards members of the Jewish movement (for example, Herod in 12.1-5 or the owners of the Philippian soothsayer in 16.19-24). Implicit in both 'sets' of characters is clothing. There are times when clothing becomes an explicit narrative actor as in the execution of Stephen above or, later, when it is ripped from its wearers – in the case of Paul and his companions in 16.22 – to make them naked and humiliate them publicly. Clothing, sometimes explicit, generally implicit and unnoticed, is taken for granted by the narrator and by the audience listening to Acts. Its presumed presence indicates the role which Earth plays, however unnoticed, for the characters and in their conduct throughout Acts.

Conclusion

Our listening to Luke's use of $g\bar{e}$-language and imagery in the Stephen story reveals a very complex ecological picture. Earth is a participant in the events that surround the formation of God's people revealed through the central characters of Abraham and Moses. Earth allows the power of God to be revealed. Over it the key figures of Israel's story of salvation travel. It is the arena of human activity and habitation.

The primary figure in Stephen's account of Israel's story is God. Only God has the power over $g\bar{e}$ to move, guide and relocate the principal figures of Israel's story. Earth's connection to creation and humanity also means that it has an agency that is divinely given. But Luke also perceives that Earth is also linked closely in the network of interrelationship between human beings and their environment. Humans have power over it; they can influence it positively or negatively. The attitude with which $g\bar{e}$ is viewed by humans means that it can become an agent of liberation or a victim of abuse and object of exploitation.

---

[25] *AEC*, p. 249.
[26] On the symbolism of the various gestures Acts' characters perform with their garments, see Brice C. Jones, 'The Meaning of the Phrase "And the Witnesses Laid Down Their Cloaks" in Acts 7.58,' *Expository Times* 123 (2011), pp. 113-8.

Luke's story of *gē* in the Stephen narrative carries other possibilities for contemporary disciples concerned about the environment. Luke identifies in this part of Acts a potential utilitarian regard for Earth. If Earth, according to Acts 7.6, can be influenced negatively by those seeking the oppression of Luke's household (and this is reflected in the trial and subsequent martyrdom of Stephen), Earth can also be influenced positively. This opens the possibility of a synergetic or symbiotic relationship between Earth and humanity that critiques a hierarchical cosmology inherited by Luke. The evangelist, though, offers a more integrative communion between Heaven and Earth celebrated in the gospel and the opening chapter of Acts. Luke's conviction of harmony between these two once-separate cosmological spheres that bifurcated the sacred from the human allows us to affirm Earth as a cooperative agent and collaborator in the story of liberation. This conviction is supported in Acts 7.7, 33 and 49, and affirmed by the gospel and Lk. 2.14.

Finally, as I have noted, Saul is introduced at the end of the Stephen story. After his conversion, he will become the principal bearer, model and expression of Luke's urban-based, evangelizing strategy. Paul's preaching and his journeys around the Mediterranean communicate more than Luke's Christology or the writer's conviction of God's ongoing fidelity. They express Luke's orthopraxis for a generation of disciples located in the Greco-Roman city speaking a voice about the relevance of a Jewish Jesus in a very different social, chronological and cultural setting. What happens after Stephen's death and the role which Saul assumes, symbolized by the honour accorded him through the garments of Stephen's killers placed at his feet, will be the focus in the next chapter.

Part Two

6

Acts 8.1b–9.31. Water and Earth

After Stephen is martyred and Paul introduced, Luke notes how the suffering endured by Stephen now takes on grander proportions (8.1b-3). A 'great persecution' unfolds against the Jerusalem Jesus followers who disperse throughout Judea and beyond, to Samaria. This dispersal flags a second geographical development in Acts. While the Jesus movement begins to move beyond Jerusalem, the apostles remain in the Holy City as faithful witnesses to the risen Jesus. Stephen is devoutly buried and Saul continues to persecute Jesus disciples – men and women – imprisoning them (8.3).

In typical Lukan fashion, persecution and suffering prepare for a scene that demonstrates the ongoing life of the suffering Jesus community that continues to grow and expand, this time into Samaria (8.1-3). This geographical expansion anticipates Luke's theological agenda, already seen in Stephen's speech. The Jesus movement is going to move beyond Judaism eventually into the wider Greco-Roman world inhabited by the Gentiles. Another one of the seven, Philip, will commence this theological and geographical expansion. His preaching in a Samaritan 'city' (8.8), accompanied by miraculous healings and exorcisms, is well received and accompanied by joy (*chara*): a gospel theme linked to God's delightful action associated with the presence, mission and ministry of Jesus (Lk. 1.14, 44; 2.10; 6.23; 8.13; 10.17; 15.7, 10; 24.41, 52). In Philip, that divine encounter of delight and joy continues.

Philip, the Ethiopian Court Official, and Water (8.4-40)

Significantly, the 'city' to which Philip comes is symbolically an ecologically related nodal collecting point for seed and other Earth products gleaned from surrounding farms and shaped by the network of human beings and Earth's goods upon which they rely. There is 'much joy' in it (8.8).

The fruit of God's presence reflected in Luke's gospel and continued through the ministry of Philip brings about the conversion of the magician, Simon (8.9-13), and continues to sustain the earthly community. The fructifying result of Philip's mission to the Samaritans is endorsed by the prompt response of the Jerusalem leaders through Peter and John. These two pray over the newly baptized Samaritans who receive the Holy Spirit (8.14-17). This becomes a second Pentecost moment in the development of Luke's geographical strategy and indicates the growth of the Jesus movement beyond Jerusalem and Judea into Samaria. The Samaritan reception of the fruit-bearing Spirit validates this next stage in the Lukan narrative and the mission to the Samaritans continues through the apostolic preaching (8.24). The journey towards 'Earth's end' now clearly commences.

Philip's role in the expansion of the Jesus movement beyond Judea is not yet complete. In a scene replete with ecological imagery, he is led along a desert road that runs from Jerusalem to Gaza. This setting is reminiscent of two other desert settings in the gospel. The first is Jesus' encounter with Satan at the commencement of his public ministry (Lk. 4.13). Here the evangelist articulates three ecological principles that assist the auditor for interpreting the stance of Jesus, Earth's Child, to Earth's goods and the ascetical spirit which must guide the disciples.[1] The second is Luke's miraculous feeding story in the city of Bethsaida which, significantly, is in a desert (Lk. 9.1-17). Through this story the evangelist critiques the irresponsible economic and depersonalising tendency to usurp and neglect Earth's fruits, and affirms the centrality of Eucharist in the re-established Jesus household.[2]

When these insights are transferred from the gospel to this third desert story in Acts 8, three echoes from the gospel reverberate: (a) the function of the word-deed which is fruitful in its encounter; (b) the role of the desert in bringing forth unexpected life; and (c) the response to the word-deed that leads to an action that satisfies.[3]

In Acts:

(a) Philip meets an Ethiopian eunuch treasurer riding in his chariot from Jerusalem back to his own country. He reads a section of the prophet Isaiah as he travels but doesn't understand the meaning of the text. In a scene similar to Luke's story of the two on the road to Emmaus (Lk. 24.25-35), the accompanying stranger needs to interpret the meaning of the Scriptures to illuminate its implications.[4] The 'word-deed' which the eunuch reads, and Philip hears and questions, already noted in its association with Earth and its gestational power in Luke's gospel study, represents a particular appreciation which Luke has for an encounter with God's word that leads to action.[5] Philip guides the eunuch to see the link between Isaiah's text and the gospel about Jesus (8.35c).

(b) As they journey together they discover water in the desert. The way Luke describes this encounter with water is a moment of revelation: 'As they were journeying along the road they came upon some water, and the eunuch said, 'Behold, water!' (8.36). This is an encounter with Earth's primeval substance. In the gospel, Luke associates the memory of water with the waters of creation and fertility, memorialized in the large 40,000 litre water bronze basin in the Jerusalem temple's forecourt (Lk. 1.9, 22), imaged in the birth waters of the pregnant Mary (Lk. 1.42) and the Jordan water from which Jesus emerges in baptism (Lk. 3.21-22) as he encounters Heaven's opening and God's voice affirming his belovedness.[6] The eunuch's discovery of water in the desert offers the auditor a memory of Exodus and its formation of God's people. Here, in Acts, a new moment of formation occurs. This time it is one

---

[1] *AEC*, pp. 106–10.
[2] *AEC*, pp. 157–61.
[3] These echoes occur within a defined structure of Acts 8, as studied by Robert F. O'Toole, 'Philip and the Ethiopian Eunuch (Acts Viii 25–40)', *JSNT* 17 (1993), pp. 25–34.
[4] *AEC*, pp. 288–9.
[5] *AEC*, pp. 73–5.
[6] *AEC*, pp. 69–75; 102–5.

who has journeyed from Jerusalem, Earth's navel and the heart of Jewish faith. He is a royal representative of the non-Jewish, African world, and sexually mutilated.[7] The eunuch enters *into* the water, but not alone. He is explicitly accompanied by Philip. Both immerse themselves as Philip baptizes the Ethiopian official: 'both went down into the water, Philip and the eunuch, and he baptized him' (8.38b, c). Earth's primordial element now surrounds and washes both, identifying one as a new follower of Earth's Child.

(c) The word-deed becomes fruitful through Philip and the eunuch's encounter with it in the text from Isa. 53.7-8 and leads to action. The Ethiopian requests baptism. As a result, and after Philip is delivered to the Mediterranean coast under the action of the Spirit, the eunuch 'went on his way rejoicing' (8.40c). Noteworthy is the focus of the text from Isaiah. The sheep and a lamb, both creatures of Earth, one about to be slaughtered and the other silent before its shearers, become images that, in Luke's Christological agenda, are types of the crucified Jesus, denied justice and whose 'life is raised from the Earth' (8.33c). This explicit Earth-recognition which Luke's Isaiah gives in this short phrase summarizes the evangelist's theological insight into Jesus' resurrection (Lk. 24.1-12).

As noted in the study of Luke's resurrection story, there are two principal actors that resurrect Jesus.[8] One is God, present in Luke's story explicitly and implicitly. The other is Earth. Earth's womb, symbolized in the beginning of Luke's gospel in the birth story of Jesus through the manger and the cloth wrappings around Earth's Child (Lk. 2.7, 12, 16), has its parallel in the gospel's final chapter. From Earth's womb the resurrected and living Jesus emerges, released from the manger-like tomb that enclosed him and linen cloth that wrapped his dead body. Luke's story of the Ethiopian baptism summarizes this in the expression drawn from Isaiah: 'his life is raised from the Earth' (8.33c). Earth is again noted as an actor in Jesus' resurrection. It is captured in this scene in Acts, as the eunuch experiences life in the waters of baptism and continues his journey towards Ethiopia and northern Africa.

There is a final point about this important pericope that is worthy of note before moving to ponder the next scene and Saul's 'conversion' (9.1-30).[9] Luke's theological agenda of narrating the initial expansion of the Jesus movement beyond Jerusalem, Judea and Judaism is explicit here in Philip's encounter with the eunuch and his consequent baptism. The baptism of the Ethiopian flags a turning point in membership of the Jesus movement. Up until now, members came mainly from Jerusalem and Judea, though those present at Pentecost represent most of the Roman Empire, including Africans from Libya, Cyrene and Egypt (2.10). Philip's baptism of this royal

---

[7] On this last point, see Johnson, *Acts*, p. 155.
[8] *AEC*, pp. 284–5.
[9] On Paul's 'conversion', see Larry W. Hurtado, 'Convert, Apostate, or Apostle to the Nations: the "Conversion" of Paul in Recent Scholarship', *Studies in Religion* 22 (1993), pp. 273–84; Seyoon Kim, *Paul and the New Perspective: Second Thoughts on the Origin of Paul's Gospel* (Grand Rapids, MI: William B. Eerdmans Publishing Co., 2002), pp. 7–12; but especially see Mark D. Nanos and Magnus Zetterholm (eds), *Paul Within Judaism: Restoring the First Century Context to the Apostle* (Minneapolis: Fortress Press, 2015).

eunuch official from Ethiopia, perhaps a 'God-fearer' returning from worshipping in Jerusalem's temple, anticipates the universality of membership in the Jesus movement, irrespective of cultural and ethnic background and sexual status. This inclusive membership will be formalized in Peter's embrace and baptism of the Roman centurion, Cornelius, and his family at Caesarea in Acts 10.

Saul, 'The Way', and Foot Travel (9.1-31)

Before this new direction in the Jesus movement formally occurs, Luke now narrates the most transformative moment in Saul's life that prepares him for the missionary function that he will play in the gradual and expansive geographical growth of the Jesus movement to 'Earth's end'. Saul travels to Damascus intent on capturing Jesus members, people of 'The Way', and returning them to Jerusalem for execution (9.1-2).

Illustration 1 Funerary stele depicting cart with rider and passengers, first to second century CE.[10]

---

[10] In the Archaeological Museum of Thessaloniki. Photo by author.

This term that Luke uses for members of the Jesus movement, 'The Way' (9.2; 19.9, 23; 22.4), expresses the nature of Jesus disciples. They are men and women on a journey, implicitly missionary oriented as outlined in Luke's schema in the beginning of Acts (1.8). They travelled, usually on foot, along paths uncommon and well-worn as they moved preaching the message throughout the Greco-Roman Empire, a point now taken up reflecting on Saul's mode of travel and before moving to look at his Damascus experience in detail. In the context of this consideration of Jesus disciples as people of 'The Way', Saul, his companions and all Jesus disciples journeyed overland by foot. This is suggested in Saul's trek to Damascus.[11] For Saul this will not be the only time that he travels overland, either implied or explicit, in Luke's story of Saul's travels over lands that surround the Mediterranean, particularly Syria, Asia Minor, Macedonia and Greece.

Saul and Luke's other characters, like Peter in 9.32–11.18, would have usually undertaken their travels walking.[12] A combination of the foot travels Saul mentions in his letters, with the overland journeys he undertakes in Acts, leads to the reckoning that Saul walked around 2,900 kilometres.[13] That he walked everywhere he went on land is apparent in his journey to Damascus as he encounters the divine light, an episode examined soon and one that is told two other times in Acts. In the repetitions of the story, the auditor learns how Saul falls to the ground (9.4; 22.7; 26.13), is commanded to stand on his feet (22.10; 26.16) and is led by the hand to complete his 240-kilometre journey to Damascus (9.8; 22.11). The language that surrounds Luke's description of what happens to Saul on the way to Damascus presumes that he is travelling by foot and not by animal as most artistic depictions have him.[14] Though camels, horses and donkeys were transport means widespread in the ancient world for people of means and status (Illustration 1), no mention is made of them in the Second Testament and never presumed by Luke in the portrait of Saul in Acts.[15] Later, in 20.13, the auditor learns that Saul (now 'Paul') undertook his journey from Troas to Assos by foot (*pezeuō*), a distance of 35 kilometres.

If the auditor imagines that most of Saul's land travel occurred by foot, this has many implications, especially from an Earth-perspective. Luke's character would have been aware of the environment in which he moved, the contours of land and the effects of weather patterns on such overland journeys; 25 to 35 kilometres would be the usual distance covered in a day's foot travel.[16] This further meant that the traveller would rely on hospitality from others if the journey exceeded this distance. The other possibility

---

[11] Helpful for a study of Paul's physical travels by foot and ship is Brian Rapske, 'Chapter 1: Acts, Travel and Shipwreck,' *The Book of Acts in its First Century Setting, Volume 2: Graeco-Roman Setting*, eds David W. J. Gill and Conrad Gempf (Grand Rapids, Michigan: William B. Eerdmans Publishing Company, 1994), pp. 1–48.

[12] Jerome Murphy-O'Connor, 'Travelling Conditions in the First Century: on the Road and on the Sea with St. Paul,' *Bible Review* 1 (1985), p. 40.

[13] Murphy-O'Connor, 'Travelling Conditions,' p. 41.

[14] For a further study of modes of overland travel, usually by foot, see Nigel Crowther, 'Visiting the Olympic Games in Ancient Greece: Travel and Conditions for Athletes and Spectators,' *The International Journal of the History of Sport* 18 (2001), pp. 39–41.

[15] Rapske, 'Travel,' pp. 9–11.

[16] Rapske, 'Travel,' p. 6.

for Saul was refuge in an inn which had its own problems: haven for thieves, accommodation with animals, and the ever-present menace of bedbugs![17] Roads were usually rough and dangerous, including those that were constructed by the Romans, though perhaps less so.[18] Earth's presence and gifts through food, drink and movement across its tracks and roads would have brought an environmental awareness that would have been uppermost in the mind of any ancient traveller, including Saul.

Saul himself, writing in 2 Corinthians, speaks of the dangers he encountered including the difficulties of his overland journey. This happened, 'through many a sleepless night, hungry and thirsty, often without food, cold and naked' (2 Cor. 11.27).

Murphy-O'Connor reflects on the journeys and difficulties which the historical Paul endured, as described in 2 Corinthians. This gives us some idea of the kinds of environmental situations in which Luke imagines his Paul would have found himself during his physically demanding overland treks:

> If Paul says that he was 'in hunger and thirst, often without food, in cold and exposure' (2 Corinthians 11.27), it is obvious that on occasions he found himself far from human habitation at nightfall. He may have failed to reach shelter because of weather conditions; an unusually hot day may have sapped his endurance; mountain passes may have been blocked by unseasonably early or late snowfalls; spring floods may have made sections of the road impassable (he claims to have been 'in danger from rivers') (2 Corinthians 11.26).[19]

Saul's Damascus Experience (9.1-19)

With this imagined appreciation of the kinds of foot journeys Paul would have undertaken throughout Acts, the auditor returns to Luke's story of Saul nearing Damascus. Near Damascus a heavenly light flashes around Saul. He is surrounded by a theophany described in astral, cosmic and meteorological imagery. Luke's expression for the light 'surrounding' Saul (*periestraptō*) incorporates the Greek root word for 'lightning' (*astrapē*) and 'star' (*aster*). This is an experience of heavenly-divine origin that brings Saul, literally, to the Earth.

The Heaven-Earth dimensions identified before in the gospel and earlier chapters of Acts combine and interrelate here as Saul encounters the heavenly Child of Earth. Saul falls to Earth (9.4a) and hears the heavenly voice, the voice of Jesus, identifying himself with those whom Saul pursues and intends to execute (9.5-6). He is instructed to rise, enter Damascus where he will learn what he is to do. When he rises he discovers his eyes are open but he is blind. He has eyes with which to see, but does not see. Saul's companions travelling with him hear the heavenly voice (9.8b) but, like Saul, see nothing. They lead Saul into Damascus.

---

[17] Murphy-O'Connor, 'Travelling Conditions,' pp. 41, 45.
[18] Rapske, 'Travel,' p. 39. See also, Murphy-O'Connor, 'Travelling Conditions,' pp. 43–7.
[19] Murphy-O'Connor, 'Travelling Conditions,' p. 41.

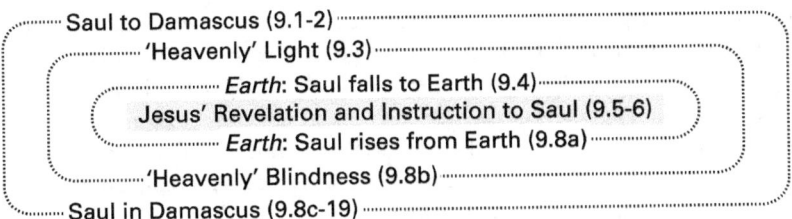

Figure 7 Saul's encounter with Heaven's Earth Child (Acts 9.1-19).

The deliberate narrative pattern that Luke creates to focus on Jesus' revelation and injunction to Saul (Figure 7) also highlights the function of Earth as a frame in which this revelation occurs.

The rest of Saul's Damascus experience involves his meeting Ananias who lays his hands on him to regain his sight, baptizes him, and encourages him to take food for strength, a simple detail to which I shall return below. Saul's visual restoration is not just physical. It is also a spiritual reversal. Those whom he sought to persecute he now stands with and begins to preach about Jesus in Damascus and amongst the Jews (9.19-22). An assassination plot against him fails as he is secreted out of Damascus (9.23-25), visits the Jerusalem disciples (9.26), meets the Jerusalem apostolic leadership group through the agency of Barnabas and 'preaches boldly' (9.27-29a). Lastly, when a final assassination attempt from the Hellenists – those with whom he was earlier aligned in their execution of Stephen – seems certain (9.29b-c), 'the brothers' bring him down to Caesarea and eventually send him back to Tarsus (9.30). Luke concludes the stories of Stephen and Saul with an upbeat summary that notes the geographical advance of the Jesus movement, now described as 'the church': 'So the church throughout the whole of Judea, Galilee and Samaria had peace and was built up, walked in the fear of the Lord and by the comfort of the Holy Spirit, was multiplied' (9.31).

Conclusion

This summary in 9.31 concludes an important story in Acts, especially given the central function which the future Saul will have in the second half of Acts. Surprisingly, Earth plays a key role in introducing Saul to Heaven's Child of Earth and bringing him finally into communion with members of the Jesus movement, confirmed through the action of God's agent, Ananias. Saul's initial blindness, a metaphor for his need for enlightenment, gives way to embracing the light of the Holy Spirit (9.17). This deep enlightenment brings him to his commitment to Jesus through baptism, an act noted in the previous scene with Philip's baptism of the Ethiopian court official. Water connects the two stories. After Saul's baptism, Luke simply writes, 'and taking food, he was strengthened' (9.19).

The food that strengthens links us back to the original moment of Saul's encounter with the heavenly voice. This fortifying food comes from the same Earth upon which Saul fell and later rose. For the auditor of Acts, this also relates to the theme of food throughout the gospel[20] and the earlier summary in Acts of what sustains the Jesus movement in Jerusalem (2.42, 46). These images allude to the Eucharistic food that strengthens, even though, for one commentator, Saul's taking of food flags his 'return to profane existence' after 'his sacral state he was in since his vision'.[21] Whether the reference to food is an allusion to the Eucharist or it moves Saul back to the world of reality that he is to face, the auditor knows that Saul's future mission has already been declared in the words of the Risen Jesus to Ananias: 'He is a chosen instrument of mine to carry my name before the Gentiles and kings and the children of Israel; for I shall show him how much he is to suffer for the sake of my name' (9.15-16). This divine script will shape Luke's portrait of Saul/Paul in Acts. In this mission, Saul will be accompanied by human companions and Earth's presence as he walks.

---

[20] *AEC*, pp. 80, 106–8, 119, 123–4, 134–5, 158–60, 163, 187, 190, 212, 217–8, 282, 300.
[21] Johnson, *Acts*, p. 165.

7

Acts 9.32–11.18. Earth's Linen Sheet

After Luke's stories of Stephen's baptism of the Ethiopian official (8.26-40) and Saul's admission into the Jesus movement, accompanied by preaching about Jesus that surprises his Damascus audience (9.1-25), Peter reappears. His appearance is significant. He is Luke's key endorsing figure for the apostolic community when a new moment is about to unfold. I have noted the gradual geographical move of the disciples beyond Jerusalem and Judea. As noted, in the Jerusalem Pentecost event, Peter interprets the meaning of the Spirit's presence for the crowd gathered in Jerusalem, anticipating its giftedness to 'all those who are far off, those whom the Lord our God calls' (2.39). Also noted in the previous chapter is the reception of the Spirit's gift in the baptism of the Ethiopian eunuch. This event suggests that God's plan is not dependent on Peter's endorsement but preparatory for what Peter will ratify for the Jesus movement. All this anticipates the fact that the moment for that authorization has arrived.

Two Healing Events (9.32-35, 36-43)

Two healing events first take place (9.32-35, 36-43). These involve Peter as Jesus' healing agent and occur at places close to ('Lydda') or on ('Joppa') the Mediterranean, away from Jerusalem, and symbolically closer to the Gentile world.[1] The geographical markers for both healings again remind the auditor attuned to the ecological hermeneutic that guides this commentary, that location is not only a site identifier, but a reminder of the wider connection which the story of Earth's children has with the wider world; a network of interrelated beings and materials that compose the space in which Luke's story is played out.

Both stories, the healing of the paralyzed and bedridden Aeneas in Lydda, and a healing in Joppa that brings life to a widow, Tabitha also called Dorcas, parallel earlier gospel events (Lk. 5.17-26; 7.11-16) and the First Testament healings by Elijah and Elisha (1 Kgs 17.17-24; 2 Kgs 4.32-37).[2] Resurrection language is central in both as Peter instructs the infirmed paralytic and the dead widow to 'be resurrected' (9.34, 40). Their response is immediate. The paralytic 'resurrects' from his bed (9.34); the widow

---

[1] Johnson, *Acts*, p. 179. Lydda, the city of Lod in the OT (1 Chron. 8.12), is in Judea (1 Macc. 11.34), en route from Jerusalem to Joppa and 17 kilometres from Joppa.
[2] *AEC*, pp. 133, 142.

opens her eyes, and sits up (9.40). Then, in a beautiful narrative touch, Luke has Peter, 'giving her his hand and resurrected her. Calling the saints and the widows, he presented her alive' (9.41).

These miracles in Acts, as in the gospels, are reminders that the power of Heaven's Earth Child and the manifestation of God's 'reign' (*basileia*) are still active through the Spirit and the apostles prominently represented through Peter. Auditors of Acts are again reminded that the restoration to human wholeness evident in these two stories and symbolized in the language of resurrection is an ecological deed. It affects all creation so that the manifestation of the *basileia*, continuing now through Earth's children, is a realization of *shalom* for all creation. In Peter's restoration to life of Dorcas, two ecological images suggest themselves.

First, she is known for her skill in tunic and garment making (9.39). She is an acknowledged artisan of textiles, Earth's products, which her widow companions display to Peter, presumably with pride tinged with sadness. This happens in an 'upper room', a memory to the Lukan auditor of Jesus' final days, as he gathers with his disciples to offer them encouragement and sustenance in the face of his impending execution. The 'upper room' is also a feminine space, one in which it is unusual for a male to be present. But with its association with the 'upper room' of Jesus' final inclusive meal in which male and female disciples participated, it is also all-encompassing, human and non-human as noted in the ecological study of Luke's gospel.[3]

Dorcas' room is implicitly gestational and fructifying given the association of textiles with the female world, textile making and wool spinning, and Dorcas' skill in crafting garments. Widows, women unable to realize their own gestational potential, weep for one who has, through her fabric working of cloth, used an important gift of Earth necessary for human beings. This reflection on the connection between cloth and Earth links back to Luke's Christology about Earth's Child, born and wrapped in cloth in an early chapter of the gospel (Lk. 2.7), and wrapped in a linen shroud towards the end of the gospel (Lk. 23.53).[4] Between these two moments, Jesus wore garments that came from Earth. Dorcas is one of Earth's children who makes such garments. She embodies, by her skilful care of Earth's gifts, a similar relationship which Luke's Jesus has with Earth in his care of Earth through his healing acts.

Besides the implicit link with the natural world which the contemporary auditor can make with Dorcas' story, there is a second ecological connection that comes from Dorcas' name. Her Greek name means 'gazelle'. As John Chrysostom comments, her name matched her character 'as active and wakeful was she as an antelope'.[5] The audience, alert to this association with Dorcas' name which Luke prefers rather than Tabitha (9.39), would know about the prominence of the gazelle in Israel's hill country and on the coastal plains around Joppa.[6]

More symbolically, the animal represented marginality. It inhabited the margins, between desert and town, the rural and urban. Considered both wild and tame, some

---

[3] See *AEC*, pp. 269-71.
[4] See *AEC*, pp. 81-3, 281, 283-4.
[5] John Chrysostom, *Homily 21 on Acts of the Apostles*, as quoted by Rick Strelan, 'Tabitha: the Gazelle of Joppa (Acts 9:36-41),' *BTB* 39 (2009), pp. 77-86.
[6] Strelan, 'Tabitha,' p. 78.

rabbis regarded it with a certain degree of holiness that, when eaten, did not compromise the purity restrictions in Jewish dietary laws.[7] This animal-association with Dorcas, if remembered, underscores the realization that the action of God, anticipated in Luke's Jesus through word and deed, is inclusive of all Earth's creatures, human and non-human. The image of the gazelle associated with the personhood and skill of Dorcas confirms that the *basileia* revealed through Peter's act is also an *ecotopia*.[8]

Cornelius and Peter (10.1-48)

With Acts' story now firmly located on the Mediterranean coast at Joppa, the scene next switches to Caesarea and the Roman centurion, Cornelius, in an episode that will permanently and definitively shift the membership of the hitherto exclusive Jewish-Jesus movement to one that will embrace the whole Greco-Roman world.[9] What now happens to Cornelius and his family, ethnically and culturally unclean in Jewish eyes, will create a dynamic that will become formally endorsed later in Acts in a council of the Jesus Jerusalem leadership (15.1-29) and confirm the direction of Saul's mission in Asia Minor. Surprisingly, the animal world will be central to this movement.

The episode begins as Luke describes Cornelius and his household as devout 'god-fearers', generous and prayerful (10.1-2). A vision, common in Acts and indicative of God's participation and initiative in what takes place, informs Cornelius to send messengers to look for Peter and bring him to Caesarea. There are several aspects about this vision. The initiative is from God. What now transpires is divinely instituted. God tells Cornelius (and reminds Luke's audience who know from 9.43) that Peter is 'lodging with Simon, a tanner, whose house is by the seaside' (10.6). Luke's purpose in giving detail of Peter's companion and location, that Simon is a tanner and whose house is 'by the seaside', is not a mere story filler. It introduces an issue with which Peter will be involved, concerned and defensive about for the remainder of his presence in Acts: the embrace of those considered impure.

A tanner was one whose primary occupation was most Earth-related. It was with animal skins regarded as unclean or impure.[10] Further, the location of the house was not for scenic reasons but to use the sea breezes to remove the odour that emerged from curing the skins.[11] Peter is with someone who is essentially impure. The odour that resulted from Simon's tanning process would have penetrated Peter's clothing. His

---

[7] Strelan, 'Tabitha,' p. 79.
[8] *AEC*, pp. 11, 130, 133, 146.
[9] On the role of the Roman military in Luke's corpus, see Gerd Lüdemann, *The Acts of the Apostles: What Really Happened in the Earliest Days of the Church* (New York: Prometheus Books, 2005), pp. 143–4. About the historicity and nature of the Italian cohort represented by Cornelius' presence, see Craig S. Keener, *Acts: an Exegetical Commentary*, Volume 2 (Grand Rapids, MI: Baker Academic, 2013), pp. 1737–42.
[10] For the technical details of tanning in the ancient world, its abhorrent smell and the use of urine for the removal of hair on animal hides, see Greg Stanton, 'Accommodation for Paul's Entourage,' *NovT* 60 (2018), p. 233, footnote 11.
[11] John Pilch, *Visions and Healing in the Acts of the Apostles: how the Early Believers Experienced God* (Collegeville, MN: Liturgical Press, 2004), pp. 88–9.

purity status that he would be expected to uphold by Jewish Jesus followers was now compromised. As John Pilch suggests,

> Peter apparently saw no contradiction between associating with this unclean fellow-member of the house of Israel, yet refusing to eat unclean food. Worse, Peter was about to discover that while he associated with an unclean fellow-member of the house of Israel, he was still unwilling to associate with non-Israelites who believed in Jesus because non-Israelites were unclean people (see Gal 2:11-14). Only a trance experience of divine provenance could bring such a 'believer' as Peter to his senses.[12]

However, Peter's disposition towards Simon prepares him for what will confront him as he goes to Caesarea and enters Cornelius' house, the home of a Gentile Roman military officer. This will be Peter's first encounter with a non-Jew whom Peter will confirm later as an authentic member of the Jesus household. Peter's action will expose the Jewish Jesus movement to a revolutionary change in direction. It will now open its door to the whole Gentile, Greco-Roman world.

Before this happens, however, God must act. This materializes through a heavenly vision that Peter experiences which will force him to re-evaluate what pertains to purity. While on the flat rooftop of Simon's house in midday prayer, like a good and faithful pious Jew, Peter becomes hungry and desires something to eat.[13] In this state he experiences 'ecstasy' and the vision (10.10).

> [11]He beheld Heaven (*ouranos*) being opened and coming down, a vessel, something like a great linen sheet (*othonē*) being lowered by its four corners upon Earth (*gē*). [12]In it were all kinds of four-footed creatures and reptiles of Earth (*gē*) and birds of Heaven (*ouranos*). [13]Then he heard a voice saying, 'Get up, Peter; kill and eat.' [14]But Peter said, 'By no means, Lord; for I have never eaten anything that is profane or unclean.' [15]The voice said to him again, a second time, 'What God has made clean, you must not call profane.' [16]This happened three times, and the thing was suddenly taken up to Heaven.
>
> 10.11-16

The vision is central to Acts. Clothing and an implicit animal presence in the gazelle-Dorcas story earlier, become explicit in Peter's vision on Simon's rooftop. Earth images, gifts and creatures are unambiguous.

Peter's vision, to which I shall return shortly in some detail, is the centrepiece between what happens to Cornelius, first in his vision to send messengers to bring Peter to Caesarea, and then, later, as Peter meets Cornelius and his household and speaks God's word intended for them (10.33). Peter's words finally lead to a second Pentecost event, as the Holy Spirit,

---

[12] Pilch, *Visions*, p. 89.
[13] Keener suggests that Peter's prayer is not at a set time for prayer and that the location of this event (on a rooftop in daylight) emphasizes the public nature of what takes place: Keener, *Acts 2*, p. 1762.

fell upon all who heard the word. ⁴⁵The circumcized believers who had come with Peter were astounded that the gift of the Holy Spirit had been poured out even on the Gentiles, ⁴⁶for they heard them speaking in tongues and extolling God. Then Peter said, ⁴⁷'Can anyone withhold the water for baptizing these people who have received the Holy Spirit just as we have?'

10.44-47

The action of the Holy Spirit and the water baptism replicates for these newly initiated Gentile converts the same event that occurred to Jesus in the gospel's early chapters.[14] The presence of the generative and fecund Spirit and the plunging into Earth's most primordial element, evoke ecological images that accompany the Gentiles as they begin to live out their discipleship in Jesus as Earth's new children.[15]

Peter Reports to the Jerusalem Leaders (11.1-18)

The action of God's spirit, as at the first Pentecost, and the baptizing initiative of Peter cements the new direction of the Jesus movement. The Gentiles are now formally incorporated as members of the group of disciples. This becomes confirmed as Peter returns to Jerusalem to justify this new divinely endorsed initiative (11.1-18). He recounts to the Jerusalem group what happened. He repeats his experience of the vision as verification for his inclusion of Gentiles into the Jesus movement, confirmed through the action of the Holy Spirit which led him to baptize Cornelius and his household:

⁵I myself was in the city of Joppa and I saw in ecstasy a vision coming down, a vessel, something like a great linen sheet (*othonē*) being lowered by its four corners from the Heaven (*ouranos*); and it came close to me. ⁶As I looked at it closely I saw four-footed creatures of Earth (*gē*), and wild beasts, and reptiles and birds of Heaven (*ouranos*).⁷ I also heard a voice saying to me, 'Get up, Peter; kill and eat.' ⁸But I replied, 'By no means, Lord; for nothing profane or unclean has ever entered my mouth.' ⁹But a second time the voice answered from Heaven, 'What God has made clean, you must not call profane.' ¹⁰This happened three times; then everything was pulled up again to Heaven.

11.5-10

In the original experience Luke mentions how Peter is 'inwardly perplexed' (10.17) and 'ponders' the vision (10.19). It is clear that the vision's meaning is not obvious. Its clarity only emerges as Peter encounters the household of 'unclean' Gentile believers. He becomes more convinced about this as he witnesses the gospel story of Jesus to Cornelius (10.34-43), recognizes the action of the Holy Spirit (10.44-46), senses the

---

[14] See *AEC*, pp. 102-4.
[15] For a discussion on the fecundity of the Holy Spirit, see *AEC*, pp. 67-9. On the primordial nature of water, see *AEC*, pp. 69-71.

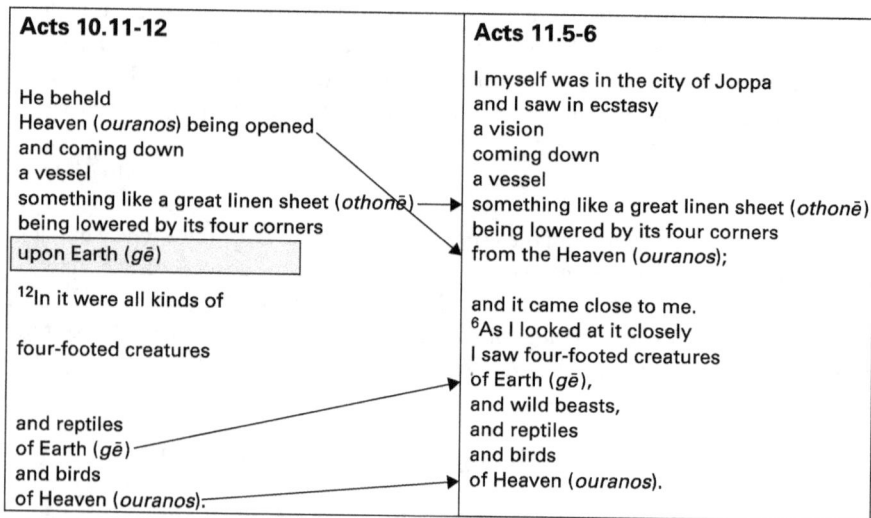

Figure 8 The thematic comparison of Peter's vision, as narrated in Acts 10.11-12 and Acts 11.5-6.

urge to baptize Cornelius and his household (10.47-48) and attests later to his Jerusalem colleagues of God's gift of the Spirit upon the Gentiles (11.1-18). He comes to the firm conviction of God's ethnic impartiality (10.34; 11.12) representative of the divine will. Peter concludes his Jerusalem witness with this conviction: 'If then God gave them the same gift that he gave us when we believed in the Lord Jesus Christ, who was I that I could hinder God?' (11.17). This statement represents the final resolution of his earlier perplexity about the vision's meaning. Central to everything is the role and function which the vision plays in Peter's experience and his later retelling of it in Acts 11. Luke's description of the original vision and Peter's recall are similar though with notable alterations.

Four things emerge from a close study and comparison of Luke's account of Peter's experience of the sheet (10.11-12) and his recount of it to his Jerusalem colleagues (11.5-6) (Figure 8): The repetition of key ecological images (Heaven, Earth, linen sheet, reptiles, birds, four-footed creatures), an emphasis on the specific ecological expression 'Earth' (gē), the communion between Heaven and Earth, already confirmed in earlier events in Acts, and the re-evaluation of the purity status of all Earth's creatures.

In the original vision which Peter experiences in Acts 10.11-12, an event which Luke emphasizes twice (10.17, 19) – hence indicating the importance and nature of what is being revealed – Peter 'beholds' Heaven, the divine dwelling.[16] In language

---

[16] That this vision of Peter and this part of Acts is founded on a Lukan tradition incorporated into the story, is argued by Lüdemann, *Acts*, pp. 143-4.

reminiscent of the gospel's scene of Jesus' baptism (Lk. 3.21), the Heaven 'is opened'. The passive nature of the verbal form indicates divine action which now repeats a theme familiar from the gospel: the deliberately orchestrated communion that results from Jesus' presence and mission.

In Luke's scene of Jesus' baptism, he is at prayer as the Holy Spirit descends upon him from the Heaven, 'like a dove'. The somatic association of an Earth-creature from the opened Heavens with Jesus affirms that communion between Heaven and Earth's Child, who possesses God's spirit and acts to bring healing to all Earth's creatures. It is worth repeating here what the Franciscan friar Bonaventure (1217–74) said about this event in Luke's gospel as he quoted the reflections of John Chrysostom (*c*. 347–407):[17] 'Now the Holy Spirit descends on [Jesus] in the form of a *dove* because of its signification, because, as Chrysostom says, 'this bird above all others is the cultivator of greatest love'. Thus, the Holy Spirit appeared to Christ in *full animal form*.'[18] The bird metaphor, an image of a creature of Earth, reveals the presence of God's Spirit descending upon Jesus.

The evangelist repeats the image of communion between Earth and Heaven in Acts symbolized by the 'great linen sheet' (*othonē*). This comes down from Heaven onto Earth. The sheet acts as a receptacle of all kinds of Earth creatures.[19] Peter considers these creatures as unclean in coherence with the cultural and religious markers that distinguished Jews from the nations (Lev. 11) and especially Hellenistic groups (1 Macc. 1.62-63).[20] Loyalty to these distinguishing purity markers and *kashrut* laws through refusal to eat certain foods (e.g.: Dan. 1.16-18; Tob. 1.10-13) was central.[21]

It would seem that these creatures are of heavenly origin and are placed on Earth through this divine action of lowering the large four cornered linen sheet, a 'vessel' (a term referring to a culinary receptacle[22]) that embraces all Earth's creatures of divine origin. This vessel is universal. The thrice repeated divine judgement on these creatures is that they are not 'common' or 'unclean' (10.14) because of God's action with and upon them (10.15).

Luke's chiastic arrangement earlier in Acts 10.14-15 (Figure 9) as Peter first experiences the heavenly linen sheet and its creatures (a – profane; b – unclean; b¹ – made clean; a¹ – not called profane)[23] underscores the tension that occurs in Peter's mind and the central issue with which he wrestles. The heavenly voice tells Peter to kill and eat the animals that have come down from Heaven in the linen sheet. Peter responds, 'By no means, Lord; for I have never eaten anything that is,

---

[17] See *AEC*, pp. 102–5.
[18] Robert Karris, *Works of St. Bonaventure: Commentary on the Gospel of Luke. Chapters 1–8* (Saint Bonaventure, NY: Franciscan Institute Publications, 2001), p. 270. Emphasis original to translated text.
[19] The 'Earth' relatedness of the creatures which the large sheeted vessel contains is de-emphasized in the NRSV translation.
[20] On the 'mixture' of animals, see Keener, *Acts* 2, pp. 1768–9.
[21] James D. G. Dunn, *The Acts of the Apostles* (London: Epworth Press, 1996), p. 137.
[22] Keener, *Acts* 2, p. 1767.
[23] Pilch, *Visions*, p. 88.

```
     (a) profane
         (b) or unclean.'
             ¹⁵The voice said to him again, a second time, 'What God has
         (b¹) made clean,
     (a¹) you must not call profane.'                                    10.14-15
```
Figure 9 The chiastic structure of Acts 10.14-15.

The resolution of this tension later will determine the unfolding mission in the rest of Acts: what and who defines purity/impurity and therefore who or what is the ministerial focus of the Jesus movement. Further, Luke structures the whole episode of Acts 10.9-23 as a call narrative along the lines of the First Testament commissioning narratives.[24]

Though there are minor differences, Peter's later description of the vision in Acts 11.5-6 (Figure 8) echoes the originating experience and emphasizes similar themes: the communion between Heaven and Earth, in fact, the juxtaposition of Earth-Heaven; the divine action involved in the lowering, the universality of the vessel that contains Earth's creatures, and the vessel described as a 'great linen sheet' (*othonē*).

The 'Great Linen Sheet' ('*othonē*') (10.11)

Though *othonē* can refer to ancient tablecloths, something more significant lies behind the term. To recover this significance, we must return to Luke's gospel and see how clothing is central for Luke's Christology.

Cloth surrounds Jesus at his birth as it identifies him as Earth's Child.[25] In Lk. 24, after Joseph of Arimathea wraps the body of the crucified Jesus in a linen cloth and lays him in a rock-hewn tomb (Lk. 23.50-53), the women go to the tomb early the next morning bringing spices to perfume the corpse (24.1). They are perplexed to see the tomb's entrance stone rolled away (24.3-4). On entering the tomb they are amazed to encounter two men in dazzling apparel who announce to them the meaning of the tomb's emptiness: Jesus has been raised (24.4-5). They then return to announce the Easter message 'to the eleven and all the others' (24.9). Their message is not believed (24.11). In response, 'Peter rising ran to the tomb and stooped to see the linen cloths (*othonē*) by themselves; and went home, amazed at what had happened.' (Lk. 24.12)[26].

The application of an ecological awareness here in Luke's resurrection narrative, and especially as Peter goes to check out the women's story, contributes to the function of *othonē* in Peter's later vision in Acts. Two Earth elements make their appearance in Luke's resurrection story: rock and cloth (*othonē*).

---

[24] Keener, *Acts* 2, p. 1761.
[25] See *AEC*, pp. 81-3.
[26] On the Lukan authenticity of this verse rather than a later interpolation, see Bruce Metzger, *A Textual Commentary on the Greek New Testament* (London: United Bible Societies, 1975), p. 184.

Peter runs and encounters the 'tomb' first, an Earth structure explicitly identified as made of 'cut stone' (Gk. *laxeutos*; Lk. 23.53). Peter's 'stooping' to observe the tomb's contents (24.12) is a gesture of ecological reverence for Earth's element. This is similar to the action of the women earlier in the resurrection story as they bow their faces to Earth in their encounter of the two men in the tomb (24.5b).[27] As Peter looks into the rock-hewn structure he sees the second of Earth's gifts, the linen burial cloths. These are the *othonē* that will thematically recur in his vision in Acts. They no longer wrap Jesus' body, but are by themselves. Given the elite status of linen in the ancient world, they become the most prominent identifiers of Jesus' eternal, regal and exalted status, a status that will become transparent in the gospel's closing scenes.[28] Peter encounters only rock and the *othonē*, nothing else. Unlike what happens in Acts 10, as God tells Peter the meaning of the *othonē*, there is no angelic appearance or divine voice clarifying for Peter the meaning of Earth's elements: the empty tomb or the *othonē*. There is no need. Earth's gifts communicate to Peter the meaning of the tomb's emptiness; Peter's encounter with Earth becomes a moment of Easter faith. As a result, Peter 'went home, amazed (*thaumazō*) at what had happened' (Lk. 24.12c). Peter's amazement (*thaumazō*) is the fruit of his seeing. This brings about a response of awe to the presence of what he perceives. It is the reality to which the stone tomb and the *othonē* point: the fact of Jesus' resurrection.

In the classical Greek world the root meaning for 'being amazed' (*thaumazō*) is often associated with an encounter with the deity.[29] Peter's amazement is not about confusion or perplexity, which will find its echo in him in Acts as he again encounters the *othonē*. In Acts the *othonē* leaves him originally perplexed and pondering its meaning with all that it contains. Later this clarifies to move him to embrace God's creatures, human and non-human, even creatures considered impure, which in Acts have a divine origin and to Heaven they return. Here in the gospel the *othonē* and stone, Earth's gifts, bring him to know the meaning of Jesus' resurrection. These, and only these, are present to him; nothing else has intervened or illuminated him. As seen in the gospel's next scene, and given Luke's narrative logic, this engagement with Earth's gifts is equivalent to an encounter with the Risen Jesus. They become the medium of this revelation. When the two return to Jerusalem from Emmaus and recount their meeting with the Risen Jesus in the breaking of bread, they learn, 'The Lord has risen indeed, and has appeared to Simon!' (Lk. 24.34). The only 'appearance' to Peter that the auditor knows about is the appearance of the tomb and the *othonē*. This means that Earth's gifts – stone and cloth – have become the means of announcing to Peter Jesus' resurrection.

The Linen Sheet's Creatures (10.12)

Coming back to the appearance of the *othonē* in Peter's vision in Acts 10, the *othonē* leaves him first perplexed, pondering the sheet's meaning with all that it contains.

---

[27] *AEC*, p. 288.
[28] On the importance of linen in the ancient world see *AEC*, pp. 82-3, 216.
[29] See θαυμάζω in *EDNT* 2, pp. 134-5.

Later, this clarifies to move him to embrace God's creatures, human and non-human, even those creatures regarded as unclean.

From the perspective of an ecological hermeneutic, there is another animal considered most unclean to Peter's Jewish colleagues that appears in the gospel, and anticipates the work of Peter's sheet animals. This is the pig. It makes its appearance twice, and both times it brings about human healing.

The first occurs in Jesus' healing of a man possessed (Lk. 8.26-39).[30] In the exorcizing process, evil spirits enter a herd of pigs who fall into the sea and drown. The act of these animals allows the man to come to full healing. The second occurs in Jesus' parable of the two lost sons (Lk. 15.11-32).[31] The younger son who has squandered his inheritance finds himself in a time of famine looking for work. In his dire situation he hires himself out to look after pigs. Through his communion with Earth's pigs, the young lad comes to his senses, realizes his wrongdoing and returns to his father's house where he is warmly received back. These gospel pigs, like Peter's unclean animals in Acts, are agents of liberation and healing, possessing divine value.[32] All the unclean creatures that appear in Peter's *othonē* are of divine origin. They come from Heaven and to Heaven they return. The pigs in the gospel and the creatures on Peter's heavenly sheet redefine what is clean and capable of interacting with humanity to bring liberation. The path to the unclean is open.

These creatures become the metaphoric representatives of the unclean and common, and the means by which God communicates the divine intention of what and who are to be purified. Animals in the Lukan corpus have a value beyond mere usefulness for human observation or consumption.[33] The dove in Jesus' baptism asserts the action of God's Spirit. The pigs in the gospel assume salvific importance. The creatures in Peter's sheet vision become the means that God uses for Peter to break beyond the stereotypical judgement of who is clean and thus belongs to the Jesus movement. Peter's reflection on these sheet creatures will later bear fruit in a direction that will define the future of the Jesus movement.

In Acts 10 and 11, Peter's vision of Earth's gift, the *othonē*, affirms God's communion with Earth and the divine origin of all Earth's creatures. Everything, human and non-human, even unclean creatures, reveals a sanctity which Peter can no longer dismiss. This flows over to his response to those 'unclean' Gentiles of Cornelius' household seeking baptism. His response also represents a most radical reaction that extends the meaning of the earlier baptism by Philip of the Ethiopian (8.26-40) and anticipates Peter's Gentile ministry.

At stake now is not only a geographical extension of the Jesus movement, as indicated in the episode with Philip, but a demographic, ethnic and cultural extension

---

[30] For the discussion on this story in the gospel, see *AEC*, pp. 155-6.
[31] *AEC*, p. 212.
[32] *AEC*, pp. 155-6, 212.
[33] For a positive study on the way Acts 10 reclaims the centrality of animals and their relationship to humanity, see Hannah M. Strømmen, *Biblical Animality after Jacques Derrida* (Atlanta: SBL Press, 2018), pp. 67–89, especially p. 72. Also, Walter Houston, 'What was the Meaning of Classifying Animals as Clean and Unclean', *Animals on the Agenda: Questions about Animals for Theology and Ethics*, eds Andrew Linzey and Dorothy Yamamoto (London: SCM Press Ltd, 1998), pp. 18–24.

that crosses the defined purity boundaries. Peter's eating a meal with Gentiles represents a radical shift in identifying God's holy community. When Peter comes to Jerusalem to explain his actions, he meets solid criticism by the upholders of Jewish Torah and purity conduct codes (11.2-3). The Jerusalem gathering of Jesus disciples, especially those from the 'circumcision party' critical of Peter's action, listen to his experience of the Holy Spirit at work amongst Cornelius' Gentile household (11.15-16). Peter concludes his testimony, 'Therefore, if God gave to them the same gift as to us when we believed in the Lord Jesus Christ, who am I to hinder God?' (11.17). Peter's witness brings his critics to silence. A conversion of heart takes place, as they recognize God's action in what has taken place through Peter. Their response concludes this section of Acts and affirms Peter's stance: 'Then to the Gentiles God has given repentance that leads to life.' (11.18c). The path is now set for a vibrant mission to the rest of the Greco-Roman world and an expansion of the Jesus movement beyond Judaism. Saul will become the principal agent for this development.

Conclusion

This section of Acts is a turning point in Luke's narrative agenda. Peter again enters the story to confirm the new direction for the Jesus movement, a movement towards the Gentiles. This turn to the Gentiles will define the next stage in the story of Earth's children that will now unfold. Two healing stories prepare for Peter's encounter with Cornelius' household and their baptism, the ritual event that cements the incorporation of this Gentile Roman household into the household of Jesus' disciples. No longer, from this moment, will the Jesus movement be purely Jewish, but a blend of Jews and Gentiles that will occur not without its tensions that will need its resolution later in Acts, especially on how to ensure that purity rituals are respected.

From this emerges the conclusion that, by the gospel's end, there has been a shift in the way creation is perceived. Early in the gospel, Earth's elements, the manger and cloth bands, identify Jesus as its child. Throughout the gospel, Earth's elements and gifts participate in his ministry. They enable reconciliation and healing. In Acts 10 and 11, Earth's gift, *othonē*, makes a second appearance to Peter in a vision that affirms God's communion with Earth and the divine origin of all Earth's creatures, including, surprisingly, the Gentiles.

Here in Acts, a vessel-like linen sheet again acts as an agent of identity. It moves Peter in a direction that he never anticipated. It will bring the gospel to the Gentiles and universally expand the household of Jesus disciples. The eternal nature of linen, first identified in the gospel, continues to be the symbolic image of life and eternity for what will take place in Acts, first through Peter, then, later, through Saul.[34]

---

[34] On the eternal nature of linen, see the discussion in *AEC*, p. 83.

8

Acts 11.19–14.28. Earth's Interconnectivity and the God of Creation

The focus of Acts now briefly moves from Jerusalem, and the endorsement of the inclusion of the Gentiles into the Jesus movement (11.18), to the wider Mediterranean world, especially at Antioch (11.19-22). Barnabas is sent by the Jerusalem members to Antioch, having heard of the growing number of converts there (11.22-24). He then goes to Tarsus to find Saul and bring him back to Antioch. This becomes their preaching base as they encourage and build up the newly formed Jesus household (11.25-26). It is here that they learn from a Jerusalem prophet, Agabus, of an imminent 'great famine (*limos*) over the whole world (*oukoumenē*)' (11.28).

The 'Worldwide Famine' (*limos*) (11.28)

Whatever the historical truth of the worldwide famine at the time when Luke writes, there is something deeply symbolic about the *limos* that links to Earth's environment.[1] This *limos* represents a major ecological disaster that would affect the whole Earth and its habitants.[2] The language which Luke uses to describe the world, the *oukoumenē*, is an Earth-related expression, symbolic of the whole Roman Empire and everything that lives on it.[3] It is also the first noted issue that faces the growing group of identified Jesus disciples outside of Jerusalem, at Antioch.

The problem that faces them concerns food, or a lack of it, a consistent theme in the Lukan corpus. The *limos*, though, is not exclusively about food and its availability. It concerns the network of ecological interrelationships that is Earth and upon which human beings depend.[4] The *limos* is symbolic of Earth's suffering. As noted in the

---

[1] Famines occurred throughout the Roman Empire. Josephus records one in Palestine during a sabbatical year (47–48 CE) when Tiberius Alexander was governor (46–48 CE), in *Ant.* 20.101. There was, however, no worldwide famine. See Conzelmann, *Acts*, p. 90. But on food and grain shortages, see Bruce Winter, 'Acts and Food Shortages,' *The Book of Acts in its First Century Setting, Volume 2: Graeco-Roman Setting*, eds David W. J. Gill and Conrad Gempf (Grand Rapids, MI: William B. Eerdmans Publishing Company, 1994), pp. 59–78.
[2] See also Leonhard Goppelt, 'πεινάσω (λιμός),' *TDNT* 6, pp. 12-17.
[3] For equating *oukoumenē* with 'Earth', see Max Zerwick and Mary Grosvenor, *A Grammatical Analysis of the Greek New Testament, Volume 1, Gospels – Acts* (Rome: Biblical Institute Press, 1974), p. 388.
[4] Winter, 'Acts,' pp. 66–8.

gospel, Jesus dines with all kinds of people, breaking the boundary codes that separate groups from one another in their attempt to preserve the holiness of the Israelite community. Food, for Luke's Jesus, becomes the means of communion and the encounter with God's delightful liberality. Earth's fruits through shared food and table hospitality enable this communion.

Jesus blurs the lines of separation that defines God's holy community. What divides, separates and classifies (into clean and unclean) is overcome in Acts through the principled and endorsing ministry of Peter to the Gentile household at Caesarea. What confronts Barnabas and Saul now is the universal lack of food, of Earth's sustenance, to meet the needs of people.

There are two moments in the gospel where Luke mentions a famine (*limos*). Both uses are instructive for our understanding in Acts. In Lk. 4.25, as Jesus preaches in his hometown synagogue, he instructs his townspeople about God's largesse inclusive of foreigners revealed in Elijah's ministry to a Sidonian widow during a severe *limos*. The point that Luke's Jesus makes to his audience is that his God is not exclusive of a particular people but for all, irrespective of cultural, ethnic or religious background. The description which Luke gives of the famine is similar to its description in Acts:

> [25]Truly I say to you, there were many widows in Israel during the time of Elijah, when the Heaven was shut up for three years and six months, and there was a great famine (*limos*) over all the Earth; [26]yet Elijah was sent to none of them except to Zarephath in Sidon, to a certain widow.
>
> Lk. 4.25-26

Luke's second gospel use of *limos* comes in the parable of the two lost sons (Lk. 15.11-32). The younger, having left his father and his father's household with his inheritance, discovers himself penniless, alone, bereft of companionship at a time when a 'great famine' sweeps across the Earth (Lk. 15.14). He eventually comes to recognize his need for reconciliation and forgiveness from his father. The rest of the story and its conclusion is well known.

In both gospel instances of *limos*, the event severely affects Earth. In the first, the famine comes about when 'Heaven was shut up for three years and six months' (Lk. 4.25). The *limos* results from the conjunction between Heaven and Earth. One affects the other. In this case, the *limos* is a global, universal and cosmic event. In the second, in the parable of the two sons, the *limos* is the context for the young man to come to realize his wrongdoing. It has a humanitarian and anthropocentric influence. It affects the young man to the point that he realizes his own spiritual *limos* and seeks to do something about it.

These gospel parallels speak into the present situation with which the two disciples in Antioch must deal. At Antioch, the universality of the *limos* implies more than lack of food, but a link to what is happening cosmically and for Earth's people. They are affected by a universal and global *limos* which could lead to division, greed and an encouragement to hoard Earth's goods for survival, a problem Jesus specifically addresses in the gospel (Lk. 12.15-49) as he encourages his disciples to be generous

(Lk. 16.19-31).[5] This spirit carries over into Acts as the disciples arrange for relief (described as a 'ministry' in Acts 11.29, Gk. *diakonia*) from the famine for their Judean compatriots.

Peter's Release and Earth's Presence: Iron (12.1-11)

Luke's story now moves back to Judea and specifically Jerusalem, describing the violence enacted by Herod upon the apostolic leadership, with the death of James, and the imprisonment and subsequent miraculous release of Peter (12.1-11). In both instances, objects fashioned from Earth are used to execute and incarcerate. Herod uses a sword to kill James (12.2) and has Peter put in prison, secured with two chains between two soldiers and a guard at the door of the prison (12.6).

To contemporary ears, the sword, chains and prison door are ferrous objects. To the modern scientific observer, iron is a product of Earth and is Earth's quintessential essence.[6] It forms the Earth's outer and inner core.[7] In scientific terms, iron is a chemical element with an atomic number of 26, and Earth's most common element formed by the fusion of stars. Whether the presence of iron is explicit or implicit in Acts, the auditor attuned to ecological matters must hear a resonance, a deep resonance, with Earth, its primordial formation, space and the cosmos. As contemporary auditors to Luke's Acts, and sensitive to the environmental issues that confront us, the imperceptible presence of iron in the narrative links us to Earth's very essence and birth. It is a reminder of Earth's gestation.

In 12.1-11, Earth's very being is used to kill, control and shackle. Despite the exaggerated security deployment, God's presence subverts Herod's plan to humiliate Peter in a public setting at Passover (12.4). This presence occurs in the form of an angelic appearance to Peter which wakes him. The drama of the scene and the ecological images that bring Peter to freedom are unmistakable:

> [7]And behold an angel of the Lord appeared and a light shone in the prison. The angel tapped Peter's side, and woke him, telling him, 'Get up quickly!', and the chains fell off his hands. [8]The angel said to him, 'Gird yourself and put on your sandals.' He did so. And the angel said to him, 'Throw your cloak around you and follow me.' [9]And going out, he followed, and he did not know that what was happening through the angel was true. He thought he was seeing a vision. [10]Passing by the first guard-post, then the second, they came to the iron gate which led to the city and opened of its own accord. Passing through, they went down along one street. Immediately the angel left him. [11]Then Peter came to himself and said, 'Now

---

[5] See *AEC*, pp. 186–7.
[6] For the intimate link of iron to Earth's core and evolution, see University of Alberta, 'Scientists discover Earth's youngest banded iron formation in western China: discovery provides evidence of iron-rich seawater much later than previously thought,' *ScienceDaily*, 11 July 2018, www.sciencedaily.com/releases/2018/07/180711182731.htm (accessed 15 July 2018).
[7] John Britt, 'All about Iron,' *Ceramics Monthly* 59 (2011), pp. 14–15; Aviva Rutkin, 'Iron Rain Left Heavy Metal on Early Earth,' *New Scientist* 225 (2015), p. 14; Hugh R. Rollinson, *Early Earth Systems: a Geochemical Approach* (Malden, MA: Blackwell Pub., 2007).

> I truly know that the Lord sent his angel and rescued me from the hand of Herod and all that the Jewish people were expecting.'
>
> 12.7-11

There are several intriguing features to this story. The shining light which accompanies an angelic appearance is already familiar to Luke's auditors (Lk. 2.9). The angel wakes Peter and tells him to get up. This action, in response to the angel's instruction, immediately releases Peter's chains.[8] Then the angel gives Peter what would seem unnecessary dressing instructions. He is told to gird himself, put on his sandals and wrap a cloak around himself (12.8). The reason for the angel's explicit instructions might seem pragmatic, to disguise himself in his escape or perhaps to protect himself from the cold night air as he ventures out of his prison into the city. It is certainly an echo of the Exodus story, as the Israelites prepare to leave Egypt, with girdles around their waists, sandals on their feet and staff in hand (Exod. 12.11).

At an ecological level of intuiting the angelic instructions, Peter's dressing himself in cloak and sandals invites the auditor to recognize the power inherent in the Earth's gifts symbolized in these articles that Peter now wears.[9] They allow for his restoration back to the Jerusalem Jesus community from which he has been removed by the political authority represented in Herod.

Luke's earlier parable of the two lost sons in Luke 15 has something similar, in which a cloak and sandals make their appearance.[10] In this well-known parable, the younger lost son takes off from the family farm with his father's inheritance, even while his father is still alive, to live an indulgent life. He later comes to his senses after experiencing severe destitution. His encounter with his Earth-roots in the form of pigs and their pods brings him to desire a return to his father's household, not as a family member but a slave (Lk. 15.14-20). The father warmly welcomes him and in an unexpected reversal of fortune, restores him, not as a slave but as his son. Two signs of this restoration are the cloak that he has wrapped around his son and the shoes placed on his feet (Lk. 15.22). He is not to be a slave, but an integral and restored member of the household.

There are similarities between the characterization of the young son in this parable and the figure of Peter in this scene in Acts. Both are trapped in some way: Peter through his incarceration, the son through his experience of poverty. Both are restored to community: Peter through God's action in the presence of the angel; the son through the father, the parabolic image of God's largesse and compassion. Both are dressed in cloak and sandals as the Earth-symbols that anticipate this restoration to full community. In the Peter-story, Earth actually cooperates to affirm Peter's freedom and release him into the city. Luke notes, 'the iron gate which led to the city... opened of its own accord' (12.10).

This third object, the gate that leads from the prison to the city, explicitly identified by Luke as made of iron, and a reminder to the contemporary ecologically-attuned interpreter of Earth's origins, acts 'of its own accord'. It becomes an agent in the narrative,

---

[8] The sequence of chains falling off the imprisoned and doors miraculously opening is found in Euripides (c. 480–406 BCE) in *Bacch.* 447–8: 'The fetters from their feet self-surrendered fell; Doors, without mortal hand, unbarred themselves', as translated by Conzelmann, *Acts*, p. 94.
[9] See the earlier discussion on the Earth-symbolism associated with clothing in Chapter 5.
[10] See *AEC*, pp. 212–14.

independent of any other being, even an angel or Peter himself, to cooperate in God's great design for the liberation of all who are oppressed, flagged initially in the inaugural proclamation of Jesus in Nazareth's synagogue that defines his future ministry in the gospel (Lk. 4.18-19).

> 18a"The spirit of the Lord is upon me, because [the spirit] has anointed me,
> b to preach good news to the poor [the spirit] has sent me,
> c to proclaim to the captives *release*,
> d and to the blind, recovery of sight,
> e to send the oppressed *released*,
> 19 to proclaim the year of the Lord's favour.   Lk. 4.18-19[11]

This text from Isaiah, a Lukan compilation of Isa. 61.1 and 58.6 (and in that enigmatic order reversal from the original text) affirms the essential Christological aspect of Jesus' ministry in the gospel. The key verb (emphasized twice above) concerns 'release'. As Luke's Jesus interprets the prophetic text, his ministry is about *releasing* the poor, the captives and the oppressed. Peter's physical and literal release from Herod's prison illustrates one example of Jesus' seminal Nazareth statement. It also sums up how this personal experience of Peter can theologically confirm the 'release' that Peter has endorsed for the Gentiles. What Peter experiences through God's agency further justifies the theological and pastoral spirit out of which Peter has already acted in his declaration in Acts 11 concerning the Gentiles, when he declares to the Jerusalem leaders: 'If then God gave them the same gift that he gave us when we believed in the Lord Jesus Christ, who was I that I could hinder God?' (11.17).

This is not the last time that the theological theme of 'release' will be found in Acts. It occurs again later, more explicitly, in the release of Saul and Silas from imprisonment at Philippi (16.25-34). As Peter and, later, Saul experience physical release in which Earth's gifts play a part, so their ministry becomes one of freedom for all those they encounter and to whom they preach. In other words, the spirit of the proclamation by Earth's Child at the beginning of Luke's gospel flows over into the ongoing ministry conducted in Acts by Earth's children.

Peter's Clothing (12.8)

There is one final detail about this story that is important, again concerning Peter's clothing. As Peter emerges from the prison into the city, the angel leaves him. It is then that Peter recognizes that what he had experienced was not a vision, as he first thought, but an act of God. He declares, 'Now I truly know that the Lord sent his angel and rescued me from the hand of Herod and all that the Jewish people were expecting.'

---

[11] To echo part of footnote 27 in *AEC*, p. 110: the translation here, as in *AEC*, literalizes the Greek text, explicates and square brackets 'the spirit' as a reminder that the subject agent implied by Luke in the verbs 'anointed' and 'sent' is the Spirit. This also prevents gendering God's Spirit. For further on this, see *AEC*, p. 110, footnote 27.

(12.11). But what brings him to this conviction? A clue is found in the gospel and Peter's experience of the empty tomb (Lk. 24.12). It is something highlighted in Chapter 7.

After his death, Jesus is surrounded by Earth's gifts in burial, wrapped in a linen shroud and placed in a stone hewn tomb (Lk. 23.52). The women come to the tomb very early in the morning on the Sabbath to anoint and perfume the body. They find the tomb open, its entrance stone rolled away, and two men inside. The two address the perplexity of the women, explaining the meaning of the tomb's emptiness: 'He is not here. He has been raised!' (Lk. 24.6). The women rush from the tomb to announce the Easter message to the eleven and the rest of their companions, only to be met by disbelief (Lk. 24.9-11). Then Luke adds, 'Peter rising ran to the tomb and stooped to see the linen cloths by themselves, and went home, amazed (*thaumazō*) at what had happened.' (Lk. 24.12).

Though there is some scholarly discussion about the originality of this verse as a later insertion, its authenticity as a work of the evangelist is generally held by most.[12] If this is the case, then auditors are left to consider how Peter comes away from the tomb with 'amazement'. Luke's word for this (*thaumazō*) suggests not incomprehension – he is amazed by a lack of comprehension of what he has experienced – but an experience that is deeply spiritual and profound. He has had a thaumatological encounter and, unlike his male colleagues, 'the eleven and all the others', comes to realize the truth of the women's Easter proclamation. But what brings him to this conviction?

I noted in the previous chapter how *othonē* functioned in Peter's vision of the heavenly 'linen sheet' and contained all the unclean creatures that he is told to kill and eat (11.5-10). Peter's experience of *othonē* in Acts connected to his experience of the linen shroud that surrounded the body of Jesus in the tomb (Lk. 24.12). Cloth is the common feature that leads Peter in the gospel to come to Easter faith and, in Acts 11, to recognize the heavenly origin of all Earth's creatures. The cloth that now surrounds Peter, in his divinely orchestrated release from prison, is the constant factor, as he responds to the angel's injunction to dress and put on his sandals (12.8). Earth's gift of clothing brings Peter to his senses and the realization that God has been at work in his release.

The Suffering 'Country' (12.20)

Peter's release brings Herod's ire against those who guarded him (12.18-19) as he turns also against the people of Tyre and Sidon. They make representation to Herod for peace because, 'their country (*chora*) relied for food on the King's country' (12.20d). This simple phrase evokes two important ecological aspects: (a) the environmental interdependence in a network of relationships, with one 'country' (*chora*) dependent on another, and disciples encouraged to use their possessions for those in need to alleviate poverty and to cement friendship; (b) Luke's teaching about wealth and the socio-political Greco-Roman context of Luke-Acts of an elite regime oppressive of the poor.

---

[12] Metzger, *Textual Commentary*, p. 184.

Luke's reflection on material possessiveness accompanied by oppression of the poor surfaces frequently in the gospel and continues, though subtly, here, in Acts. As seen earlier, the Jerusalem followers of Jesus are encouraged to share their possessions to relieve those in need (4.32-37). Their witness, however, remains in a world of economic disparity and need. The expression that Luke uses in 12.20d ('their country [*chora*] relied for food . . .') indicates that the residents of Sidon or Tyre were not the only ones to suffer. Reliance on food means that the wider ecological space that linked to these towns, Luke's *chora*, ('the country'), was also deprived. It did not have the resources with which to address the physical necessities of the people that relied upon it.

Despite the brevity of its description, 12.20d underscores the lack of availability of Earth's fruits for all. Greed rather than a spirit of largesse still dominates the Greco-Roman world in which Earth's children live. This includes those new places outside of Judea and Galilee where foreigners live, symbolized by Sidon and Tyre, and to where the followers of Jesus now venture.

Jesus' parable of Lazarus and the rich man (Lk. 16.19-31) finds its regal parallel here. In the gospel parable, wealth is used to exploit the poor.[13] This happens especially through the greedy accumulation of Earth's goods, squandered by one who is royally dressed in 'purple and fine linen' (Lk. 16.19b). The wealthy person in the parable can allow Earth's goods to fall from his table because of material overabundance unable to be contained on his table (Lk. 16.21). He has enough goods to satisfy the needs of beggars like Lazarus. As the parable story comes to its resolution, death comes to the wealthy character who goes into eternal anguish; Lazarus dies and is received in the bosom of Abraham. A divinely executed 'chasm' of judgement separates the wealthy elite from the poor and creates an insurmountable reversal in the afterlife.

The elements of Luke's parable repeat themselves in what happens next in Acts (12.20-23). The regal figure of Herod, the symbolic representative of elitism, wealth and prosperity, dressed in royal robes and acknowledged as a god, 'immediately' suffers the same fate as Lazarus' wealthy oppressor (12.23a). His death represents God's judgement on one who did not acknowledge the divine source of all wealth and goods (12.23b), a theme already witnessed earlier in Acts, in the deaths of Ananias and Sapphira (5.1-10). In contrast to the death of Herod and despite the experience of famine, a divine reversal occurs, as in the Lukan parable, 'The Word of God grew and multiplied.' (12.24).

The Growth of God's Word by Barnabas and Saul (13.1-49)

From a later theological perspective, the 'Word of God' is a metaphor for the totality of God's self-revelation in Jesus.[14] Luke's Jesus is concerned about the whole of creation, including humanity. This recognition continues to flow over into Luke's second volume, as Earth's children spread the word of the Risen Jesus, and establish households of

---

[13] See *AEC*, pp. 216-8.
[14] For a fuller discussion on the Lukan function and meaning of the 'Word of God', consult *AEC*, pp. 56-9.

disciples outside of Jerusalem and Judea. God's word flourishes. In the next section of Acts, Luke narrates how two of Jesus' followers, Barnabas and Saul, bring about the growth of God's word. They are part of the disciples' gathering in Antioch (13.1-2) and then commissioned for 'God's work' (13.2-3).

In the first of three missionary journeys that Saul undertakes, the two sail first to Cyprus (13.4). Here Barnabas and Saul (whom Luke now calls 'Paul': 13.9) proclaim God's word to its inhabitants and its proconsul, Sergius Paulus, but not before meeting resistance from a local magician (13.5-12). They then move on to their next site of proclamation, sailing from Cyprus to Perge, the capital of Pamphylia, a province in Southern Anatolia (Asia Minor).

Ships and Sea Travel

At this juncture in Acts, with Luke's mention of sailing (13.4), it is important to reflect on the nature of sea travel, the means by which people in the ancient world undertook this, and the construction of vessels upon which Luke's Paul would have sailed. All these have ecological implications for the contemporary listener of Acts.

Earlier in Luke's narrative in Acts a ship would have been the main means of long-distance transport around the Mediterranean. Sometimes such travel is explicit by Luke, as in the case with Paul and Barnabas above. Other times, it is implied; for example, when one of the Seven, Stephen, travels to Phoenicia, Cyprus and Antioch preaching about Jesus (11.19-21), or when Barnabas goes from Antioch to Tarsus to bring Paul back to Antioch to support and teach the fledgling Jesus community there (11.25-26). Stephen's travel between Antioch and Cyprus could only happen by boat. Similarly, while it would be possible for Barnabas to go to Tarsus overland, the most efficient way would have been by one of the vessels that would regularly sail between the Antioch harbour of Seleucia and Tarsus. As noted above, Antioch becomes a centre of the Jesus movement under the tutelage of Barnabas and Paul. Their first missionary venture is by sea that takes them from Seleucia to Cyprus' harbour at Salamis, overland to Paphos (13.4-6) and then again by ship to Perge and to the Roman province of Pamphylia (Illustration 2, below).

The crossing from Paphos to Perge becomes the first time that Luke explicitly mentions that Paul and his entourage 'set sail' (13.13).[15] Here they begin their preaching tour throughout the provinces of Pamphylia and Lycaonia travelling to Perge, Pisidian Antioch (13.13-50), Iconium, Lystra, Derbe (14.1-7), Pisidia and Pamphylia to the harbour city of Attalia (14.24-25). From here, and for a second time, Luke explicitly mentions how Paul and Barnabas 'sail' back to Antioch from where they began their journey (14.26). Luke's next explicit mention of 'sailing' comes at the beginning of the second missionary journey as Barnabas and Mark go to Cyprus (15.39), while Paul takes Silas and travels overland through Syria and Cilicia back to the cities visited in

---

[15] On the nature and composition of those who accompanied Paul in his travels, see Stanton, 'Accommodation,' pp. 227–46. Other places where Paul is accompanied by companions and an entourage include 14.26; 15.37-38; 16.11-15, 40; 17.7; 27.1; 28.16, 30.

Acts 11.19-14.28. Earth's Interconnectivity and the God of Creation

the first journey (15.40-41). Finally, they 'sail' from Troas on the western coast of Asia Minor, across the Aegean Sea to the island of Somathrace, eventually coming to Macedonia to Neapolis, the harbour of Philippi, where the next stage of Paul's preaching activity unfolds (16.11-12).

Luke employs implicit 'travel notes' to move his hero, Paul, from one place to another, to venture across seas, without naming the means by which this happens. The presumption for the auditor is that this would happen by ship. In just three verses, in 18.21-23, for example, Luke has Paul travel from Caesarea on the Mediterranean coast, north to Antioch of Syria, then Galatia, Phrygia, most of Asia Minor, finally returning to Ephesus. In 20.5-6, Luke has Paul sail from Ephesus to Macedonia, then from the west coast of Asia Minor to Caesarea using a complex navigational route (21.1-8). Luke leaves Paul's most dramatic sea journey until the last chapters of Acts as he undertakes a final sea voyage to Rome (27.1-28.14). Paul's ultimate sea journey will be our focus in Chapter 13.

As discussed in Chapter 6, Paul's overland mode of transport would have been by foot. However, for convenience and to hasten a journey from one part of the Mediterranean to another, Paul would have travelled by ship. This would have been the case with Paul's desire to go from Miletus (western Anatolia) to Jerusalem (via Caesarea) in time for Pentecost (20.16). If the distances that Paul travelled by

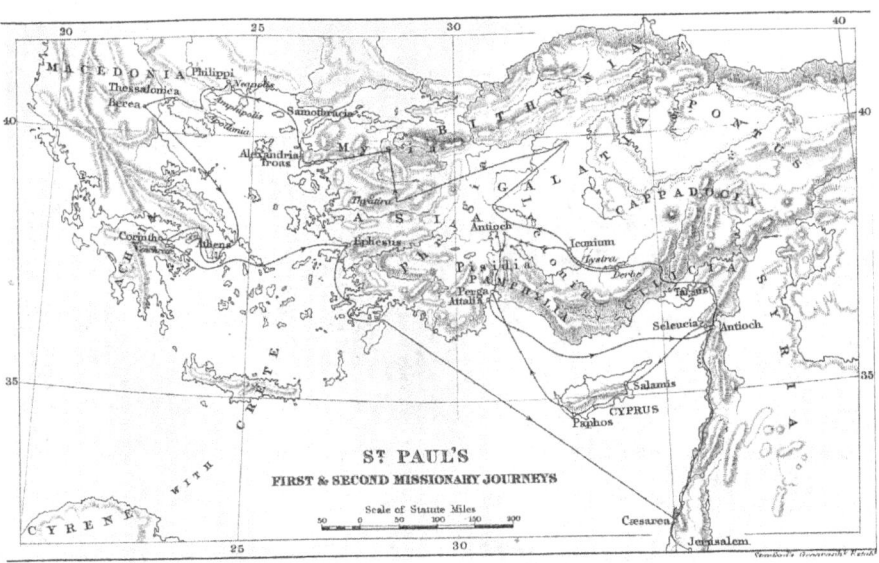

Illustration 2 Paul's first and second missionary journey.[16]

---

[16] William Smith, *Concise Dictionary of the Bible* (London: John Murray, imprint of Hodder & Stoughton, 1900), facing p. 693.

ship through all his voyages were calculated, they would amount to over 2,000 kilometres.[17]

Paul's sea travels, like those of any ancient voyager, were seasonally conditioned and potentially dangerous, as later learnt in Acts 27. From the end of May to the middle of September were considered the best times to travel by ship. Other periods were variable and mid-November to mid-March dangerous, an observation which Luke confirms in 27.9 as Paul and his entourage sail towards Italy 'after the Fast', that is after Yom Kippur, which occurred around the autumnal equinox, outside the usual sailing season given the unusual lateness of the Fast.[18]

The types of ships on which Paul travelled were not specifically designed for passengers. They were grain or cargo carrying vessels and allowed passengers only if space was available (Illustrations 3 and 4). In these circumstances, passengers were

Illustration 3 Roman ship with stowed amphorae, 75–60 BCE.[19]

---

[17] Murphy-O'Connor, 'Travelling Conditions,' p. 41.
[18] Rapske, 'Travel,' p. 22; Murphy-O'Connor, 'Travelling Conditions,' p. 45. In 59 CE – the possible year that Luke envisages Paul's voyage to Caesarea and his intention to get to Jerusalem for 'the Fast' – the date of Yom Kippur (10 Tishri) was early to late October (Rapske, 'Travel,' pp. 24–5).
[19] http://www.ancientportsantiques.com/ancient-ships/merchant-ships (accessed 4 February, 2019); drawing after the Madrague de Giens shipwreck, estimated dimensions 40 × 9 m and 3.5 m draught for 375 tonnes of cargo (by Jean-Marie Gassend, 2005), copyright ascribed to Arthur de Graauw © 2011–2019, reproduction permitted.

incidental to the precious cargo. Greed rather than hospitality dictated availability of space for passengers who had to provide for themselves, live on the deck without any cabin accommodation, sheltered by the shade of the mainsail or their own private tent, and cook meals after the crew had been fed.[20] People like Paul on such a sailing vessel in these circumstances were not going on a holiday cruise! They were vulnerable to the elements, and fear of drowning or their vessel capsizing was ever present.

The vessels that sailed across the Mediterranean were mainly grain carriers that offered great financial benefits to their owners. Their tendency would be to ensure that their ships sailed as often as they could, even outside the limits of the best sailing periods. Pliny comments on the greed of shipping owners that determined the number of trips that their vessels made: 'Not even the fury of the storms closes the sea; pirates first compelled men by the threat of death to rush into death and venture on the winter

Illustration 4 Amphorae for transport and storage of food products.[21]

---

[20] Murphy-O'Connor, 'Travelling Conditions,' p. 46.
[21] Archaeological Museum of Thessaloniki. Photo by author.

seas, but now avarice exercises the same compulsion.'[22] Cicero makes a similar observation as 'sailors are apt to hurry things with an eye to their own gain'.[23]

This background helps us to understand the situation that Paul and his companions later find themselves in as they experience a storm and become shipwrecked *en route* to Italy in Acts 27.13-44. Greed, the vice that commandeers Earth's goods for personal gain with a disregard for others, and against which Jesus taught his disciples (Lk. 12.13-21), determines the overall shipping schedules of clippers around the Mediterranean. Lucian (125–80 CE), writing in the second century, indicated that the minimum annual income from one grain carrier, the *Isis*, was 72,000 drachmae. This would reflect values similar to cargo vessels in Luke's day not many decades earlier.[24] One drachma was the equivalent of a day's wages for a skilled worker in the ancient world.[25] A consideration of the size and construction of the vessels themselves also leads to ecological considerations.

If Lucian's description of the *Isis*, written well after Luke completed Acts, reflects even remotely the kind of grain-carrying ships of Luke's day and upon which Paul

Illustration 5 An inscribed funerary stele of a sailor in his small vessel.[26]

[22] Pliny, *HN*, 2.47.125, as quoted by Rapske, 'Travel,' p. 28.
[23] Cic., *Fam.* 16.9.1, as quoted by Rapske, 'Travel,' p. 28.
[24] Lucian, *Navigium (The Ship or the Wishes)* 13.
[25] Michael Vickers, 'Golden Greece: Relative Values, Minae, and Temple Inventories,' *American Journal of Archaeology* 94 (1990), p. 613.
[26] Archaeological Museum of Thessaloniki. Photo by author.

travelled, these vessels were large.[27] They were huge constructions from Earth's fabric, metal and wood – this included softwoods like pine, fir, and cedar[28] – and beautifully decorated. Lucian writes of the *Isis*,

> What a huge ship! A hundred and twenty cubits long [55 metres], the ship-wright said, and well over a quarter as wide [14 metres], and from deck to bottom, where it is deepest, in the bilge, twenty-nine [13 metres]. Then, what a tall mast, what a yard to carry! What a fore-stay to hold it up! How gently the poop curves up, with a little golden goose below! And correspondingly at the opposite end, the prow juts right out in front, with figures of the goddess, Isis, after whom the ship is named, on either side. And the other decorations, the paintings and the topsail blazing like fire, anchors in front of them, and capstans, and windlasses, and the cabins on the poop – all very wonderful to me. You could put the number of sailors at an army of soldiers. She was said to carry corn enough to feed all Attica for a year.[29]

While there is evidence of smaller vessels (Illustration 5),[30] the overall impression from the nautical information and marine archaeology of this period reinforces the impression about the great labour force and large quantities of Earth's resources needed to construct these vessels. These were not ecologically neutral vessels. They were floating conservatoria of Earth's natural assets. Auditors would keep this in mind when they pick up Luke's story of Paul's other voyages in the remaining chapters of Acts. But to return, now, where I left Paul and Barnabas before embarking on this study of ships and sea travel, as they journey to Pamphylia.

The First Missionary Journey of Paul and Barnabas (13.4–14.26)

The pattern or rhythm which Luke establishes in the opening scenes of Barnabas and Paul's first missionary journey – of proclamation, success and rejection, and final advancement of God's word – will repeat itself over most of the remaining chapters in Acts. At the heart of this pattern is the lived reality of Luke's gospel meta-parable of the sowing seed (Lk. 8.4-21). Like the parabolic seed, despite initial frustration, the seed of God's word will flourish through the preaching of Barnabas and Paul, and, later, by Paul himself. This Lukan pattern is evident when Paul and Barnabas preach in the synagogue of Pisidian Antioch (13.13-47). Their eventual rejection becomes the reason for Barnabas and Paul's definitive 'turn' to the Gentiles, away from their Jewish audience.

With this story before us, of Paul's first missionary journey, two points can be made. The first has already been made in Chapter 4. There, Luke's anti-Jewish bias, which first

---

[27] See the summary analysis of grain carrying ships and shipping in the ancient world by Nicole Hirschfeld, 'The Ship of Saint Paul, Pt 1: Historical Background,' *BA* 53 (1990), pp. 25–30.
[28] Michael Fitzgerald, 'The Ship of Saint Paul, Pt 2: Comparative Archaeology,' *BA* 53 (1990), p. 35.
[29] Lucian, *Navigium*, 5 (Loeb Classical Library). The *Isis* would have weighed 1,114 tonnes (Rapske, 'Travel,' p. 31).
[30] Fitzgerald, 'The Ship,' p. 37.

surfaced in Peter's Pentecost speech (3.12-26), is clear. This theme recurs here in the address which Paul and Barnabas make in the synagogue of Pisidian Antioch. Their conviction leads to a missionary principle which Luke's main protagonist, Paul, will articulate in later scenes, especially in 18.6 and 28.28.

It is worth repeating Luke's concluding words to this event in Pisidian Antioch. They remind contemporary auditors of this important and influential writing, the ongoing potential which such an unreflected bias can have on interreligious dialogue today. This is especially so in the wake of the rapprochement between Christianity and Judaism in recent decades, the recovery by Christians of their Jewish roots, and their recognition of a history of Christian anti-Judaism that has fostered hatred and contributed towards the extermination of Jews in modern history.

For Luke, the rejection of Paul and Barnabas in the Pisidian synagogue occurs because of Jewish jealousy and revilement of their speakers (13.45). They respond in words that sum up Luke's missionary strategy: 'The Word of God had of necessity to be spoken first to you. Since you reject it and do not judge yourselves worthy of eternal life, behold we turn to the Gentiles' (13.46).

The next verse brings us to our second point. Paul and Barnabas acknowledge that the justification for this move to the Gentiles comes from the Word of God, as they interpret Isa. 49.6, a text to which Luke has already implicitly referred back in 1.8d: 'For since the Lord commanded us, "I have appointed you to be a light to the nations, that you might bring salvation to the end of the Earth."' (13.47). They consider this biblical word the grounds for their decision to turn to the Gentiles. The Lukan characters interpret the text here in terms of their non-Jewish mission. At another level, and as discussed in Chapter 2, the original text of Isaiah referred to God's action in bringing light to the nations. Humanity becomes the means by which this light of salvation reaches Earth's 'end': 'I will give you as a light to the nations, that my salvation may reach to the end of the Earth' (Isa. 49.6c).

In Acts 13.47 Luke changes the original meaning of Isaiah to make the apostolic missionaries the immediate and direct harbingers of God's salvation. In other words, Luke's reinterpretation of Isaiah seeks to elevate the status of Paul and Barnabas as God's agents of salvation to the Gentiles. Despite this Lukan tweaking of the original prophetic text, its alteration and perhaps because of it, this salvation – God's action and healing implied in the theological motif of salvation – is intended to 'reach to the end of the Earth'. From this phrase in Chapter 2, it is clear that this includes more than a group of human beings resident in the wider Greco-Roman world, living outside Jerusalem and beyond Judea. Earth itself is the subject of the salvific action of God's agents, a theme that permeates the gospel through the ministry of Earth's Child. This agency continues into the story of Acts through Earth's children, in this case through Barnabas and Paul. Earth is an ecologically related interconnectivity of relationships that concerns every organism on the planet, including humans, Jews and Gentiles. While our natural focus is on those who are influenced by this Word and become believers, there are also ecological implications and consequences that flow from people's responses to Jesus' Word and the action of the Spirit which imbues the new Jesus households. The Gentiles delight in their inclusion in this new missionary focus from Paul and Barnabas (13.48).

However, the wider ecological sense included in the expression 'the end of the Earth' is also present: 'The word of the Lord spread abroad throughout the whole of the region (*chora*)'. (13.49). *Chora* refers to 'country' or 'rural precinct'.[31] It is one of the many Earth images that compose Luke's linguistic environmental taxonomy. The word of the Lord is not exclusively anthropocentric, but embraces the whole 'region' into which Barnabas and Paul are now sent as God's agents of salvation.

The God of Creation (14.1-28)

As they travel to other cities in the region, they again meet resistance and hostility from their co-religionists in Iconium, Lystra and Derbe, cities of Lycaonia (14.1-6). Their fidelity to their mission brings healing to a crippled man in Lystra. In response to the adulation that is heaped on them for the healing and the perception that the locals have of them as Greek gods, Paul and Barnabas seek to correct the perception of the Lystrans.

The speech which Luke places on the lips of the two is one of the most explicit theological statements in Acts of the nature of the God of creation. It is a compressed ecologically oriented expression of the Creator God active in the healing deeds of God's children represented by Paul and Barnabas. They want their audience to know that it is not their innate godly power that brought the man healing, but the action of their God:

> [15]We are human beings like yourselves, bringing you the good news to turn away from vain things to a living God who made the Heaven and Earth and the sea and everything that is in them, [16]who in past generations allowed all the nations to walk in their own ways, [17]although God did not leave God's self without evidence, showing kindness (*agathourgeō*), giving you rains from Heaven and fruitful seasons, filling your hearts with food and gladness.
>
> 14.15-17

Paul and Barnabas urge their listeners to recognize that the source of everything is God. Their God lives and is active (v. 15b, above). The healing of the crippled man is evidence of this. But the God they witness to is the creator of everything: of the cosmic spheres, Heaven and Earth, and the primordial waters of creation, 'the sea' (v. 15c).[32] God's creative act then continues into the present through the ministry of Paul and Barnabas. They remind their audience that the evidence of God's presence is shown by God's 'kindness' (*agathourgeō*) revealed in gifts from Heaven ('rains') and Earth ('fruitful seasons').

The ecological largesse of their God brings gladness to the hearts of human beings. The implications of this are unmistakable. The human encounter with creation enables each person to come to know the God which Paul and Barnabas proclaim. The text which Luke draws into these verses (v. 15b) when the two refer to the living God ('God

---

[31] See the discussion on *chora* in *AEC*, p. 296, footnote 3.
[32] For this 'sea', see *AEC*, pp. 73, 98, 103, 131.

who made the Heaven and Earth and the sea and everything that is in them') comes from Exod. 20.11. The context for the reference in Exodus is the giving of the covenant. The 'ten words' of the covenant are preceded with the reminder of the Creator God and God's action in the story of creation in Genesis 1, 'In six days the Lord made Heaven and Earth, the sea, and all that is in them, but rested the seventh day; therefore the Lord blessed the sabbath day and consecrated it.' (Exod. 20.11, NRSV).

The identification of the God of 'Heaven and Earth, the sea, and all that is in them' (Exod. 20.11a) is a preamble to the reason for the Sabbath rest, the day on which God 'rests' after six days of creative activity. This divine day of 'rest' consecrates the Sabbath, the seventh day, for the Hebrews. Given the implied non-Jewish audience addressed by Luke's two missioners, this reference to the consecration of the Sabbath is dropped. Nevertheless, the whole context of the original setting for this Exodus text and Luke's later incorporation of it – albeit with adaptation for a different implied audience unfamiliar with or unaccustomed to the spirit of the Sabbath – the ecological connectedness and environmental association with the God of Heaven and Earth are retained and still evident.

This first journey of Paul and Barnabas concludes as they retrace their steps through this part of Asia Minor, revisiting Derbe, Lystra, and Iconium, offering encouragement and support, but not without suffering persecution from their antagonists from Antioch and Iconium (14.19-23). After sailing from Attalia, they finally return to Antioch from where they first launched their preaching tour. They assemble the Jesus followers of Antioch, report what had happened on their journeys and how God 'had opened a door of faith to the Gentiles' (14.27c).

Conclusion

This section of Acts brings Paul on to Luke's missionary stage. First introduced by Barnabas, a respected apostle from Jerusalem, Paul accompanies Barnabas and is second named by the author, but gradually assumes a leadership role in the first of three major journeys that Paul will undertake in spreading the word to 'the end of the Earth'. Luke's agenda is to cement the spread of the Jesus movement, originally Jewish, beyond Judaism and Judea, to the wider Greco-Roman world. As observed earlier, the manner or pattern in which this 'turn' to the Gentiles occurs, as Barnabas and Paul preach, involves rejection, and sometimes persecution, wins supporters or has some success as Luke notes the ultimate growth of God's word despite apparent setbacks. The most explicit declaration of this move to the Gentiles occurs in Acts 13: 'The Word of God had of necessity to be spoken first to you. Since you reject it and do not judge yourselves worthy of eternal life, behold we turn to the Gentiles.' (13.46).

This pronouncement sets the agenda for the rest of Acts, unfortunately, as already noted, supported by the author's undercurrent of anti-Jewish sentiment: the Jews have rejected God's invitation to join the Jesus movement; it is to the Gentiles that the invitation will now be given. While Luke's implicit Gentile audience celebrate the new direction, indicated above and in Chapter 4, the result of this Lukan anti-Jewish presentation, coupled with other NT texts about the 'Jews', has had unfortunate and sad consequences for the Jews in subsequent Christian history.

The declaration to turn to the Gentiles is also anticipated by earlier events that lead to it: Philip's baptism of the Ethiopian (8.26-40), Peter's healing activity outside of Jerusalem, near or on the Mediterranean coast (9.32-43), his heavenly vision of the sheet with the heavenly injunction to reconsider everything as divinely cleansed (10.9-16), then his baptism of Cornelius' Roman household that formally and definitively incorporates the first non-Jewish, Gentile family into the Jesus movement (10.17-48).

This action, one of inclusivity and release, is first anticipated by the gift of God's spirit coming upon Peter's Gentile audience (10.44-47). It is later confirmed in two ways: by the Jerusalem leaders after hearing Peter's witness (11.1-18) sanctioning his action; and, in a more personal but theologically symbolic and ecologically metaphoric way, by his physical release from Herod's imprisonment by an angel, as Peter dresses himself with Earth's clothing of cloak and sandals for what awaits (12.1-19). The Gentile missionary agenda is explicit in Acts here. Luke also incorporates into the narrative notes that have important ecological resonances.

The Earth on which the Word of God spreads is affected by its inhabitants and those who use its goods. It suffers a major ecological event, a famine (*limos*) that is universal (11.28). The ecological suffering that results is not limited to one moment in Acts. Later, Luke notes how the country (*chora*) is affected, with the reliance of one area of the Mediterranean on another for food resources (12.20). The imbalance of the availability of Earth's resources, a theme common in other stories in Luke's gospel, echoes in Acts as the obvious wealth of Herod's lands – implicitly the result of greed – serves to create dependency and want in other parts of the Mediterranean, explicitly areas that are foreign and Gentile. This confirms the reliance on Earth's goods and the network necessary for ecological sustainability. Glimpsed very briefly in one verse in Acts (12.20) is the pattern of ecological degradation so familiar in our world that creates poverty and the growing disparity between the rich and poor.

Luke continues to remind the auditor of the original vision for the Word of God which is intended to move 'to the end of the Earth'. This occurs in 13.47. This movement, in an anthropocentric sense, will embrace all peoples, Jew and Gentile. However, Earth's end also implies that Earth itself is the subject of the good news addressed by Earth's children as they continue to proclaim God's words. That is, no part of Earth, human and non-human, will be devoid of the thankfulness that arises from encountering the Word. This is the experience of 'thankfulness' which even the Gentiles learn from Paul and Barnabas as they reflect on the actions of the God of creation (14.15-17).

This brings us to a final but important ecological theme that surfaces in this section of Acts. In the speech which Paul and Barnabas address to their Lystran audience, Luke again sums up the heart of the evangelist's creation-centred theology about God. The God which Paul and Barnabas proclaim is the creator of 'Heaven and Earth and the sea and everything that is in them' (14.15). This note about God as creator of 'the sea and everything that it is in them' would have been confirmed by Paul's sea travels. Luke's Paul is a geographer, in the sense that he was geographically astute, as he moved along the major Roman highways, provincial roads and tracks, and used ships taking advantage of tides and prevailing winds. He would have been acutely aware of the environment in which he moved, the condition of seas and lands, and appreciate the God who created them.

The relationship which God has with the cosmos, the spheres of creation and the very primordial substance that represents the first act of creation ('the seas'), conveys a particular and important overriding interpretation about Earth and all that dwells on it. If God is the creator of all that exists, including Earth and the creatures of Heaven, Earth and sea ('and everything that is in them'), which Luke asserts through the theological characterization of Paul and Barnabas, then all that exists is essentially good and blessed. This will have consequences in Paul's future ministry in Acts.

Rather than encountering an essential negativity or rejection in the people and cultures which Luke's Paul will come across, the Greco-Roman world of Paul's future journeys and the Earth in which this world exists is quintessentially good and receptive to the good news. Creation will confirm Paul's message because it is already imbued with God's presence. Paul will bring the 'good news' to Earth that is already the bearer – albeit implicitly and at times without public confirmation – of goodness.

There is one hurdle that remains before Paul can embark on this journey to Earth's end. It concerns the relationship between Jews and Gentiles and how they will be able to participate communally in the Lord's Supper in a manner that does not compromise purity laws. This is a central moment of decision-making for the growing and expanding Jesus movement as it now embraces the wider Greco-Roman world. How it resolves this major issue will be the focus of the next chapter.

9

Acts 15.1–16.40. Earth Acts at Philippi

I come now to the watershed narrative moment in Luke's story of Earth's children and the mission to the Gentiles. Torah-observant Jesus disciples from Jerusalem confront the Gentile followers in Antioch. They direct that they should be circumcized according to Jewish law (15.1). After debating them, Paul and Barnabas, along with others, go to Jerusalem to sort this matter out (15.2-3). When they eventually arrive in Jerusalem, they meet the apostolic leaders and witness to God's actions among the Gentiles (15.4). They are again confronted by observant Jesus followers, members of the 'Pharisee party', who insist that new members should adhere to the injunctions of the Torah (15.5).

Illustration 6 A section of the *Via Egnatia* in Philippi.[1]

[1] Photo by author.

The Jerusalem Meeting, Paul's Second Journey Begins (15.6–16.10)

In their discernment as to how best respond to the situation, the apostles and elders listen to Peter's witness and his divinely endorsed conviction that the Gentiles who have come to faith should be treated without distinction from the original members of the Jesus household (15.6-11). Barnabas and Paul relate their experience (15.12). Finally, James, the leader of the Jerusalem apostolic community, sanctions the new direction of confirming in peace the Gentile believers and judges it best that they should adhere to certain ritual practices that would allow Jewish and Gentile Jesus followers to live together in harmony and share table communion (15.13-21).[2] James' judgement ratified, the Jerusalem leaders send a delegation along with Paul and Barnabas to Antioch to communicate the Jerusalem decision (15.22-33).

The two then remain in Antioch teaching and preaching. Paul suggests that Barnabas should return to the places that they first visited together (15.36). But after a dispute over John Mark, Paul and Barnabas part company, Barnabas sailing to Cyprus with John Mark. Paul, with Silas, returns to Pamphylia, as he commences his second missionary journey along the popular *Via Egnatia* (Illustration 6). He travels overland by way of Syria and Cilicia (15.36-40), eventually being directed by the 'Spirit of Jesus' to leave Asia Minor and travel across to Macedonia (16.1-10). They eventually arrive at the city of Philippi, a Roman colony and central city in Macedonia (16.11-12). Here, in a carefully constructed scene at Philippi in Macedonia (Figure 10, below), Luke constructs the story of Paul's meeting with Lydia, an event that is filled with rich ecological overtones.

Sabbath Rest by a River (16.11-40)

As Luke tells the story of this encounter, the evangelist uses the plural 'we' in narrating what is happening, and not for the last time.[3] The effect of the 'we' and 'us' is two-fold. It moves the auditor of Acts into a more personal engagement with what is unfolding. It also allows the author to engage the auditor directly.[4] This means that 'we' cannot sit

---

[2] Conzelmann, *Acts*, pp. 118–21; Tannehill, *Narrative Unity*, p. 190.
[3] Other 'we' passages occur at Acts 20.5-16; 21.1-18; 27.1–28.16.
[4] For those who present the 'we' passages as evidence of the historical reliability of the account from either the author-as-eyewitness or sources upon which the author depended, see C. K. Barrett, *A Critical and Exegetical Commentary on the Acts of the Apostles: Volume II, Introduction and Commentary on Acts XV–XXVII* (London: T. & T. Clark, 1998), pp. xxv–xix; also Joseph Fitzmyer, *Acts of the Apostles: a New Introduction with Introduction and Commentary* (Yale: Yale University Press, 1998), pp. 98–103. On the other hand, William S. Campbell, *The 'We' Passages in the Acts of the Apostles: the Narrator as Narrative Character* (Atlanta: Society of Biblical Literature, 2007), while not contradicting the historical implications of the 'we' passages, moves towards, in my opinion, a more fruitful approach founded on the literary strategy of characterization: 'This effect moderates the narrator's responsibility for narrative claims and, at the same time, conveys a familiarity between narrator and reader that draws readers into the story, bonds them to the first-person plural company that includes the narrator, and encourages them to evaluate positively the narrator's perspective and to accept the narrator's version of events. Indeed, even when the narrative identifies the first-person plural referents, readers experience a more personal relationship with the group and, therefore, a more sympathetic view of the narrator's perspective than is the case with third-person narration'. (p. 90).

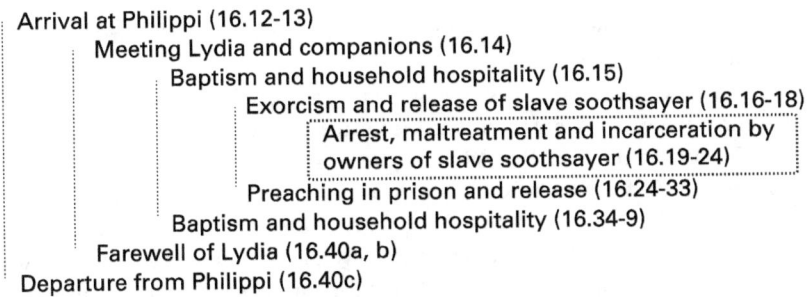

Figure 10 The chiastic structure of the Philippian Event (Acts 16.12-40).

outside the scene but, through the author whom we regard as a 'person of integrity and trustworthiness', we are invited to enter it.[5]

Listening to what happens at Philippi with 'green ears' there are several noteworthy features. The first concerns the day and the setting: 'On the day of the Sabbath, we went outside the gate beside a river where we thought prayer to be, and sitting down we spoke to the women who gathered.' (Acts 16.13). Sabbath memorializes God's rest on the seventh day, after six days of creative activity, as remembered in Genesis 1. It is an important day that surfaces frequently in Luke-Acts.[6] This day of rest, the Sabbath, celebrates the goodness of creation and the spirit of release and freedom intended through God's presence permeating creation. This links to Luke's first mention of the Sabbath. It occurs in Lk. 4.16-30 in the programmatic description of Jesus' public ministry couched in the words of LXX Isa. 61.1 and 58.6. Jesus' ministry proclaimed on the Sabbath in his home synagogue of Nazareth recognizes his mission of release and freedom that will characterize his ministry in the rest of the gospel.[7] As Jesus exercises this ministry, he is concerned with the whole of creation, human and non-human. With Paul and his companions (the 'we') gathering on the Sabbath, the auditor remembers all the other Lukan moments when Jesus acts on the Sabbath to restore creation (as, for example, in Lk. 13.10-17).[8] His deeds of healing on the Sabbath are

---

[5] Campbell, *The 'We' Passages*, p. 89. Pervo, *Acts*, pp. 392–6 offers a helpful summary of the scholarship on the 'we' passages that coheres with the position that I adopt here. Pervo concludes, 'narrative explanations illuminate the use of "we" in Acts but do not fully elucidate it. "We" is not a single character and therefore unlikely to represent the author. "Participation" in the sense of the "we" of the community (as in John 1.14), is an explanation that best gives credit to the author. This does not eliminate the possibility that the author overlooked the conflict between anonymous, omniscient narration and use of the (theoretically) limited first person and thus inadvertently created most of the problems. The use of the "we" does not identify the author of Acts. It does serve to enhance the credibility of the narrative and to associate the narrator with the person of Paul. It is a bid to be recognized as an exponent of authentic Paulinism and to authenticate the Paulinism of Acts' (*Acts*, p. 396). Stanley Porter ('Excursus: the "We" Passages,' *The Book of Acts in its First Century Setting, Volume 2: Graeco-Roman Setting*, eds David W. J. Gill and Conrad Gempf (Grand Rapids, MI: William B. Eerdmans Publishing Company, 1994), pp. 545–74) suggests that the 'we' passages are a predefined independent source, perhaps from one who was an eyewitness, which Luke, not their author, incorporates into the Acts storyline.

[6] This favoured reference for Luke, of the Sabbath, occurs 52 times in the gospels overall, with over a third of them (20) in Luke's gospel. It occurs 10 times in Acts.

[7] *AEC*, pp. 110–2.

[8] For a study of the connection of the Sabbath with creation, see Anne Elvey, *An Ecological Feminist Reading of the Gospel of Luke: a Gestational Paradigm* (Lewiston, NY: Edwin Mellen Press, 2005), p. 278, n. 242.

filled with ecological overtones intended for the liberation of all Earth's members, human and non-human.[9]

This liberating theme continues now in Acts as Paul and his friends gather at a river, an environmentally significant place, outside the urban structure that defines the city of Philippi. In this setting, on the Sabbath, there is the recognition that this is the appropriate setting for prayer. Rather than a formal place of gathering and prayer, creation, Earth's river, provides the most appropriate environment for women to gather to pray to the God of creation, acknowledged several times in Acts and particularly in the earlier address by Paul and Barnabas to the Lystrans (14.15).[10] Luke focuses on one woman in particular, Lydia: 'There was one woman named Lydia, a dealer in purple cloth from the city of Thyatira, a worshipper of God, who heard us. The Lord opened her heart to pay attention to what Paul was saying.' (16.14). Lydia and her household are then baptized. She then invites Paul and his companions to stay at her house (16.15).

Lydia seems to be a person of some wealth and a woman of independent means.[11] Her origins were from western Asia Minor. She would have been a merchant who travelled to Philippi, perhaps regularly.

The fact that she owns a considerably sized house in Philippi implies several things.[12] She would have lived in Philippi for a period of time as a residential merchant rather than as a travelling seller passing through en route to other merchandising urban centres. Lydia might even have migrated there from her original town of Thyatira in western Anatolia. Further, her status was not determined by a male. This means that she would have been either widowed, divorced or remained unmarried; she was independent; she would have known and been known by Philippian residents and integrated herself into the urban population. Her position for Luke in Acts parallels the status of Martha in the gospel (Lk. 10.39-42): the host of a household where Jesus followers gathered (16.38).[13] If host, then it is conceivable that Lydia, like Martha, would eventually act as host for the celebration of the Lord's Supper.

More pertinently to this commentary, Lydia's wealth and status came from her sale of cloth infused with a purple dye extracted from the murex shell found on the Mediterranean coast of western Asia Minor.[14] Lydia, then, was a woman whose connections were strongly linked to the availability of one of Earth's gifts considered more valuable than gold, given the process, the number of shells needed for the dye

---

[9] *AEC*, pp. 192–3.
[10] Luke alerts the auditor to the synagogue as the usual setting to gather for reading the scriptures, formal teaching and prayer, for example, in Lk. 4.16-30; Acts 13.14, 27; 15.21; 18.4.
[11] Amanda C. Miller, 'Cut from the Same Cloth: a Study of Female Patrons in Luke–Acts and the Roman Empire,' *Review & Expositor* 114 (2017), pp. 203-10. Also important for a study of Lydia, see Richard S. Ascough, *Lydia: Paul's Cosmopolitan Hostess* (Collegeville, MN.: Liturgical Press, 2009).
[12] On the size of Lydia's house, her background and occupation, also see Stanton, 'Accommodation,' pp. 234-6.
[13] *AEC*, pp. 175–8.
[14] Pliny the Elder (23–79 CE) summarizes the various processes for extracting 'Tyrian purple dye' in *HN*, 32.25-7; Lloyd B. Jensen, 'Royal Purple of Tyre,' *Journal of Near Eastern Studies* 22, no. 2 (1963), pp. 104–18, offers a comprehensive overview of the history, manufacture and cost of purple dye in the ancient Mediterranean world.

extraction and the ultimate cost of the dye in the ancient world.[15] Thus, Luke's mention of her occupation, which seems almost incidental, evokes several ecologically significant reflections. Her contact with Earth through her occupation and the beautiful linen goods she handles deepens her sensitivity to the messenger of and message about Earth's Child. She becomes Paul's first convert in Macedonia. She emulates the gift of generosity and hospitality encouraged in Jesus' disciples.

Earth's Gifts Abused (16.22)

The character or figure of Lydia frames Luke's story of Paul's first engagement with the first of many potential Jesus followers on Macedonian soil. After her baptism and offer of hospitality towards Paul and his companions (16.15), their next encounter is with a female slave soothsayer, whom Paul exorcizes (16.16-18). This results in their arrest from those who see themselves now financially disadvantaged by the woman's exorcism (16.19).

In contrast to the way Earth's gifts of clothing and shell-extracted dye can allow Lydia to come to meet Paul and be open to his message, three acts performed against the messengers of Earth's Child reveal an abuse of Earth's creation in an attempt to silence and ridicule them.

First, Paul and his companion, Silas, are dragged into the commercial centre of the city before magistrates, who, 'tearing their clothes off them ordered them to be beaten with rods (*rabdizō*)' (16.22). Here Luke offers a contrast between the kind of elegant and luxurious clothing in which Lydia dresses people and the implied nakedness that Paul and Silas now experience in 16.22. Their humiliation and beatings parallel the kind of treatment that Jesus receives in the gospel. Their public degradation occurs through nakedness. Clothing, Earth's gift, which covers them – and is implied with every other character in Acts, thus subtly underscoring the way Earth accompanies all the *dramatis personae* of Luke's work – offers protection, safety and warmth. Its removal underscores their vulnerability and the shame intended to be brought upon them by the public authorities.

Second, this shame is further compounded by the way they are 'beaten with rods' (*rabdizō*). This act has as its root the Greek word for 'wood' (*rabdos*), indicating the use of one of Earth's materials to punish and demean, a second act of Earth abuse.

Third, Paul and Silas are then thrown into inner prison, with their feet 'fastened in stocks (*xulon*)' and, presumably, as indicated later, shackled with iron chains (16.23-24).

---

[15] Jensen suggests that in Ancient Rome, the cost of an ordinary purple robe was equivalent to $US2,000; Diocletian (244–311 CE) purchased 320 grams of purple silk from Sidon for the modern equivalent of $US3,000: Jensen, 'Royal Purple,' p. 115. Jensen further attests, 'Since the largest snail contained only three of four drops of the fluid leuco-base together with the gland tissues, the processing must be done by many patient people. A recent experiment using about 12,000 *Murex brandaris* and employing good refining technique yielded 1.5 grams of crude dye. It is difficult to equate costs in antiquity with inflationary currency of today. An educated guess would equate the cost of one gram in 200 B.C. at about $[US]30.00 or $[US]13,000.00 a pound': Jensen, 'Royal Purple,' p. 109.

The Greek word, *xulon*, translated here as 'stocks', refers explicitly to wood. It is an expression applied to various wooden implements used as instruments of torture and punishment.[16] The use of wood in this case completes a triad of Earth's objects used to humiliate and shame Paul and Silas. Their use in this instance will contrast with the freedom offered by Earth's gifts in the scene that follows (16.25-34). It comes about through an act of divine intervention. In characteristic Lukan fashion, what seems to be disaster for Jesus followers soon becomes a revelation of God's power.

Release through Earth's 'Quake' (16.26)

At midnight the imprisoned duo are praying and singing hymns with their fellow prisoners listening (16.25). Their liturgical prayer leads to a cosmic, Earth-shattering event that heralds divine action and their release from incarceration: 'Suddenly there was a great earthquake ['Earth's quake'] so that the foundations of the prison were shaken; immediately all the doors were opened and everyone's chains (*desmos*) were released.' (16.26).

Significantly, Earth acts in this moment of theophanic revelation.[17] It becomes an agent in an event that will bring about freedom for Earth's children and, as I shall explicate shortly in a discussion of the verbal forms of 'shaken' and 'opened', allow the power and presence of God to enact this release of Philippi's prisoners. The earthquake ushers in God's presence. This repeats an already familiar theme in Acts: the communion between Heaven and Earth. This unity again reveals itself in the earthquake. The Earth collaborates as an agent of divine action; it enables the two incarcerated disciples to be freed. The quake rocks the very foundation of the prison and its power to constrain. The very objects, the *xulon*, doors and chains, that hinder the freedom of Earth's two children, Paul and Silas, are sabotaged by Earth's seismic action.

Further, the passive tense of the two main verbs ('shaken', 'released') indicate that God has entered the scene. God acts to bring about the release of Paul and Silas. This is signalled by two dramatic events produced by the earthquake: the opening of the doors of the jail and the dismantling of the chains (*desmos*) (16.26). To a contemporary auditor who would remember Luke's earlier story of the release of Peter from prison (12.6-11), with its similarities to what happens here with Paul and Silas, both the door and the chains are presumably ferrous objects. If this is the case in the Lukan imagination, as it could well be in the Peter episode, then these are products of Earth's resources. From the discussion about the nature and function of iron in the Peter story of Acts 12, it symbolizes the very essence of Earth and the primal fruit of its birth. The action of God restores the two companions to communion with humanity as envisaged in the Genesis narrative of creation and the divinely-imbued intention that all is 'good' that God creates and comes from Earth (Gen 1.1-2.4).

---

[16] Zerwick, *Grammatical Analysis*, pp. 405-6.
[17] On the theophanic nature of the earthquake, see Gunther Bornkamm, 'σεισμὸς, σείω,' *TDNT* 7 (Grand Rapids, MI: Wm B. Eerdmans Publishing Co., 1964), pp. 196-200.

As the metal doors of the prison swing open and the iron shackles fall from the wrists of those imprisoned, God's promise of freedom and Jesus' proclamatory declaration of his ministry, as explicated in the Nazareth synagogue in the gospel, is tangibly demonstrated a second time: first in the earlier release of Peter (12.1-11), and now with Paul and Silas. In this scene in Acts 16, Jesus' own disciples experience physical *release*: the primary verbal action in Jesus' Nazareth proclamation (Lk. 4.18c, e). They, and especially Paul, will continue to reflect Jesus' ministry in their preaching and ministry, to bring this release to Earth's end. An anticipatory symbol of this more global, universal effect of their mission is the 'earthquake', 'Earth's quake' – a cosmic, universally affective event of ecological significance – which accompanies their freedom from incarceration.

Water, Hospitality and Food (16.33-34)

What takes place next underscores their freedom to undertake this mission and how they continue to enact Jesus' Nazareth proclamation in their ministry. As the jailer is about to commit suicide thinking that his captors have escaped, Paul prevents him (16.27-29). The warden's request for salvation leads him to receive the preaching of Paul and Silas (16.30-32). The scene concludes with actions that involve ecological elements illustrative of the power of Earth's gifts to bring release, healing, salvation, faith and communion. The jailer, 'took them at that same hour of the night, and washed their wounds, and he, with all his family, was baptized immediately. Then bringing them up into his house, he set food before them; and rejoiced with his entire household having found faith in God' (16.33-34).

Water, a significant ecological symbol throughout the Lukan corpus, becomes again the symbol of cleansing and freedom here. It completes the healing which the jailer seeks to offer Paul and Silas. It becomes Earth's gift which they use to baptize the jailer and his entire household, and bring them into communion with all those who are Jesus' disciples. They are now part of a movement first established in Jerusalem, then expanded to incorporate an African (8.26-40), those from the Mediterranean coast (10.44-48), and now a Macedonian.

The domestic setting in which the meal takes place with its accompanying hospitality is infused with a sense of environmental harmony: communion through food (16.34), the precious gifts of Earth fashioned to celebrate, enjoy and sustain, leads to communion with God that brings deep joy in the hearts of those who have now 'found faith in God' (16.44). This scene in the jailer's household echoes the gospel setting of the Eucharistic meal which Jesus shares with his disciples, his final meal before his ascension and the hospitable ministry and table communion seen frequently throughout the gospel. All these gospel scenes and, now in Acts 16.34, the meal which the jailer shares with Paul and Silas, evoke ecological images of food gifted from Earth's resources and shared freely.

This combination of images, of water, household hospitality and food, contribute to the joy of the jailer and his household. This domestic assembly is not a simple gathering of parents and children. It would constitute a network of relationships, including slaves, and replicate the usual makeup of a Greco-Roman household familiar to Luke's audience. This domestic network of familial relationships parallels the ecological relationships of water and food (possibly bread and wine) that brings delight to the heart and allows human beings to rejoice, too, in the presence of God in whom they have come to believe through the actions of Earth's most primal element, water. The environmental richness of the scene is clear, though, usually, unnoticed by contemporary auditors.

This section of Acts concludes with a change of heart from the city's magistrates. They want to release the two quietly (16.35-36). Paul, on the other hand, is not keen to go quietly. He demands that the magistrates themselves come to release them (16.37). After an apology for the way they have treated these two newly identified Roman citizens, the judicial and civic leaders bring them to the outskirts of Philippi and request them 'to leave the city' (16.38-39). As they depart, the Philippian scene concludes in the same way as it began, with a visit to Lydia and her colleagues (16.40). With this, the auditor's memory goes back to the beginning of this section in Acts, to Paul's arrival at Philippi and his initial meeting with Lydia and her friends.

Conclusion

Acts 15.1–16.40 offers the auditor Luke's strategy for encountering and including into the growing Jesus movement the newly attracted Gentile believers. In Chapter 7, Peter, the leader of the Jerusalem Jesus disciples, endorsed the new moment of growth as he baptized the Roman household of Cornelius at Caesarea (10.1–11.18). Peter's divinely orchestrated release from his Jerusalem prison in the next chapter of Acts, in 12.6-11, anticipates what will happen to Paul (and Silas) at Philippi. The freedom experienced by these disciples in Luke's captivity stories continues the theme of *release*, first flagged by Jesus' synagogue proclamation at Nazareth, that continues in the gospel as an important Christological motif, and remains a critical theological motif in Acts. Not only are Jesus' disciples Earth's children, released, but, in both captivity stories, Earth's gifts and materials play an active part. Earth itself is also released.

This liberating theme carries over into Acts, first experienced by Earth's children at Pentecost. Key figures, principally Stephen, Peter and, now and for the rest of Acts, Paul, demonstrate how God's spirit of freedom revealed at Pentecost, manifests itself amongst the people and cultures beyond Judaism. The consequences of this endorsement emerge as Paul ventures forth to continue his missionary preaching in the Gentile world of Asia Minor (11.19-26; 13.1–14.28) and now in Macedonia, the focus of the present chapter.

What soon emerges in the growth of the Jesus movement, with its inclusion of the Gentiles, concerns the need to preserve communion amongst its diverse members, given their various religious and cultural identifiers, and the purity implications for the Jewish members. The decision which the Jerusalem leadership group takes, not to burden the Gentile members more than necessary, but to act in ways that would not create division, especially over purity laws and eating regulations, is a centrepiece of theological diplomacy. It establishes the foundational principle for the expansion of the Jesus movement, initially Jewish, in its embrace of non-Jewish, Gentile adherents.

According to Luke, this decision for Gentile inclusion did not come easily, especially for the more traditional and originating members of the Jesus-Jewish movement. Hints of this surface in Acts. Paul's letters, especially in his Letter to the Galatians, identify the conviction which some Jesus followers held that new members needed to be fully Judaized. The historical Paul takes a different path, recognizing the importance of Gentiles preserving their religious autonomy, free from Torah regulations imposed upon them by Israelite Jesus members. The Paul of Luke's Acts participates in a more conciliatory approach, as represented in the Jerusalem decision in Acts 15. The Gentiles will need to observe certain conditions in their communal participation and celebration of the Lord's Supper as they join with their Jewish Jesus co-religionists.

Finally, it would be important to note that in this section of Acts, Paul and his various companions travel overland through Asia Minor and cross to Macedonia by ship. These are not environmentally neutral acts. Travel entailed the use of several means which had consequences for Earth's resources. To conclude and repeat a mantra-like ecologically-oriented observation, Earth is a participant in this part of Luke's narrative. It becomes a collaborator in God's action that brings freedom for Earth's children as they continue their mission to 'the end of the Earth'. Nothing, not even nakedness, beatings and incarceration, will prevent this taking place.

10

Acts 17.1–18.1. The God of Life and Breath

Paul and Silas leave Philippi and continue their journey to Thessaloniki. The connecting road between these two cities was the *Via Egnatia* (Illustration 6), the principal Roman road constructed in the second century CE that began at Byzantium (later, Constantinople), ran through Thrace, to Dyrrachium on the west coast of Macedonia. It was a central avenue for exporting goods overland to Rome. A ship would take goods and travellers from Dyrrachium to Brindisi to link with the *Via Appia* and Rome. The 1,120 kilometre, 2.5 to 8 metre-wide stone and slab paved road, with a foundation of small stones and gravel and finished with a layer of packed earth, was a major engineering feat that linked the region (Illustration 6). Its construction relied on intense and prolonged human labour and great physical resources that had environmental consequences.[1]

Paul and Silas pass through Amphipolis and Apollonia to Thessaloniki, to a synagogue of the Jews (17.1). Luke tells us that for three Sabbaths Paul discussed with them the meaning of the scriptures about Jesus' death and resurrection and as God's 'Christos', God's anointed one (17.2-3). There is a mixed reaction to Paul's preaching. Some, Greeks and leading women, join Paul and Silas (17.4). Others, Jews described by Luke as 'jealous' (17.5) gather a rabble and attack the house in which the two are staying. They 'drag' the household owner, Jason, and other Jesus members outside before the people and city authorities. Before they finally release them (17.9) they accuse them of sedition against Rome's imperial authority. They are accused of offering their allegiance to the 'king' Jesus, rather than to Caesar, and of having 'turned the world (*oikoumēnē*) upside down' (17.6-7).

The Earth (*oikoumēnē*) 'Turned Upside Down' (17.6-15)

The Greek expression which Luke uses here, *oikoumēnē*, represents both the world inhabited by human beings and the Earth upon which humanity depends for its existence.[2] This is an ecologically loaded expression. The accusation levelled against

---

[1] Yannis Lolos, 'Via Egnatia after Egnatius: Imperial Policy and Inter-regional Contacts,' *Mediterranean Historical Review* 22 (2007), pp. 273–93.

[2] Zerwick and Grosvernor, *Grammatical Analysis*, p. 408, consider *oikoumēnē* a linguistic substitute for *gē*, Earth.

Paul, Silas and their Thessalonian companions concerns the apostles' perceived attempt at undermining Rome's imperial power. They are also accused of subverting the whole Earth itself upon which Rome and its peoples depend. This is a global or universal indictment that has ecological resonances and implications.[3]

Their antagonists believe that Paul and his companions have turned the 'Earth upside down'. There is truth in this. At one level of interpretation, especially one that is ecologically aware, the Earth, the *oikouminē*, has been 'turned upside down', not by Earth's children, but by the agency of Earth's Child and the spirit that he has imparted to his followers. A brief thought about Rome's imperial system, its hierarchical structure and the importance of 'grain' is in order at this point to recognize the ironic truth in the way these Jesus followers 'have turned Earth upside down'.

The population of the Roman Empire was about 70 million, with 90 per cent rural and 10 per cent urban.[4] The largest population, of one million, lived at Rome, the centre of power, politics, wealth and imperial presence. A diet of oil, dry legumes and especially wheat dictated what needed to flow into the Roman markets.[5] Though Italy and the network of grain fields that surrounded Rome could supply some of what was needed, wheat was rapaciously culled from Rome's conquered provinces including Macedonia. Over two thousand ships a year, carrying an average of 6,700 tonnes of grain, sailed to Rome's harbour, Ostia, from around the Mediterranean.[6]

Ostia's harbour began as a river-port on the Tiber in the mid Republican period. From the reign of Claudius however (41–54 CE) a new artificial port was constructed, known as *Portus*, just to the north of Ostia on the northern bank of the Tiber, and 24 kilometres from Rome. Trajan (98–117 CE) later expanded the harbour to make *Portus* one of the main grain and market conduits to Rome, only rivalled by Puteoli, the harbour at which Paul finally arrives on his final journey to Rome (28.13c). Thessaloniki and Neapolis ('Kavala'), the seaport of Philippi, were two of the central points of departure for ships to Rome from Macedonia and where grain and other commodities were sourced.[7] At Neapolis, 150 kilometres east of Thessaloniki, Paul and his companions first landed in Macedonia from Asia Minor (16.11).

If grain and other merchandise were commandeered from the provinces for Rome's population, this maintained and reinforced a particular perspective and control over

---

[3] In the classical Greek world, *oikoumēnē* was a geographical term that implied the inhabited world (as attested by Xenophanes 21.A.41 and Herodotus 3.114; 55.110). However, the LXX uses *oikoumēnē* around 40 times to translate the Hebrew *eretz* in the Psalms. This refers to the whole Earth, its ground, soil, natural objects and inhabitants. Luke uses this LXX Earth-perspective for *oikoumēnē* in Lk. 2.1, here in Acts 17.6, and, later in Acts 24.5. See Horst Balz, 'οἰκουμένην,' *EDNT* 2, pp. 503–4; Otto Michel, 'ἡ οἰκουμένην,' *TDNT* 5, pp. 157–9.
[4] Peter Temin, *The Roman Market Economy* (Princeton, NJ: Princeton University Press, 2012), p. 252.
[5] Temin, *Roman*, p. 30.
[6] Temin, *Roman*, p. 40. Others estimate Rome's annual consumption of grain ranged between 180–360,000 tonnes (Rapske, 'Travel,' p. 25).
[7] Livy (c. 54/69 BCE–12/17 CE) identifies the ancient Thessalonian harbour during the final Macedonia war (150–148 BCE). The harbour was large enough for Gaius Marcius to shelter his fleet there and ravage the city (Livy, *Ab Urbe Condita*, 44.10). According to Appian (*B Civ.*, 44.106), in 42 BCE Brutus and Cassius used the harbour of Neapolis for their fleets in their preparation to fight against Mark Anthony and Octavian at the battle of Philippi.

Illustration 7 Trajan's Column photographed around 1896.[8]

the *oikouminē*. Rome's program of balanced reciprocity sought to reinforce political elitism over those from poorer economic contexts on the socio-political ladder. Its rapacious manner of gleaning provincial grain and material products kept its silos full and ensured that Rome's population was satiated.

One example of Rome's rapaciousness is depicted on Trajan's triumphal column (Illustrations 7 and 8). Although the column itself was constructed and dedicated a few decades after the writing of Acts, in 113 CE, it reflects a style of Roman propaganda glorifying the empire's power to confiscate and command Earth's resources to fulfil its imperial agenda of universal conquest. The column's 190m spiral frieze, that winds

---

[8] Trajan's column, attributed to Apollodorus of Damascus (50–130 CE). Photograph: Conrad Cichorius, *Die Reliefs der Traianssäule* (Berlin: Verlag von Georg Reimer, 1896), Tafel 1: public domain, https://commons.wikimedia.org/w/index.php?curid=5074858 (accessed 7 February 2019).

Illustration 8 One of the panels on Trajan's Column depicting the Dacians deforesting the Earth prior to battle.[9]

23 times around the column with 153 panels, celebrates the victory of the Emperor Trajan (98–117 CE) in his wars against Dacia (roughly today's Romania) in 101–102 CE and 105–106 CE.[10] As depicted in the column's friezes, this victory occurs, though, not without Earth's devastation (Illustration 8).

The victory is accomplished through deforestation, the cutting of trees, the damming and diverting of rivers, building siege works, and the mining of resources. The frieze chronicles the conquest of the trees and the land along with the Dacian people. The upward spiralling of the sculptural panels, unfolding the narrative of conquest, and the explicitly religious iconography, make an eschatological claim for Rome's conquest of the whole world. Eternal life, the column suggests, is 'empire without end'.[11]

The column's depiction of the use of Earth's resources for Rome's military ends parallels the way Rome's elite also commandeered resources for their own ends. This reflects the power implanted in the social elite from their inherited socially reinforced economic script. One commentator notes,

> Socioeconomic status was patterned in a sharply hierarchical structure, with about 90 percent of the population living at or near subsistence. The superwealthy

---

[9] Cichorius, *Die Reliefs,* Tafel XLVIII (A major engagement (Scene LXVI); Dacian landscape (Scene LXVII)): public domain, https://commons.wikimedia.org (accessed 7 February 2019).

[10] For an informed visual of the column, see National Geographic, *Trajan's Column – Reading an Ancient Comic Strip,* https://www.nationalgeographic.com/trajan-column/ (accessed 7 February 2019).

[11] Barbara R. Rossing, 'Trajan's Column and the Cargo List of Rev 18:12-13: John's Critique of Rome's Economy in Ecological and Eschatological Perspective,' presentation at the *Society of Biblical Literature* meeting, San Antonio, 2016.

elite comprised the top 3 percent, while another 7 percent (approximately) were successful merchants, or owners of small properties, which allowed them to have income from rents, or owners of skilled slaves who increased profitability of workers' workshops. These merchants, landlords, and workshop owners belonged to the middle strata where subsistence was not an issue.[12]

The conventional consumption of Earth's resources by the imperial system of power, politics and control over Rome's provinces around the Mediterranean, reflected in Trajan's column, was further symbolized in the confiscation of grain and the other important provincial commodities which Rome considered essential. The place in which Luke's Paul now finds himself becomes more than a geographical location or staging post for Luke's story. It is an eco-theological symbol of the role which Earth plays in Luke's narrative, and the attitude which Jesus disciples have to Earth reflective of Jesus' teachings from the gospel.

While it is Earth's Child who has turned the *oikoumine* 'upside down', his agents are now ironically labelled as those who have reconfigured Earth. Their style of living, already identified in the earlier chapters of Acts, displays a manner that critiques Rome's imperial economic strategy. Their presence in Thessaloniki is considered corrupting of the *oikoumine* and Roman stability. There is nothing left for them but to be secreted out of Thessaloniki to Beroea, where they receive a warmer reception in the Jewish synagogue (17.10-12). However, antagonists from Thessaloniki incite crowds against Paul who is immediately ushered out of Beroea for Athens by ship (17.13-15). With Paul's relocation to the great city of Athens, a new chapter in Luke's story of Earth's children unfolds.

Paul's God of Earth and Heaven (17.16-23)

In Athens, Paul continues his engagement with its Jewish community, the Athenians gathered in the agora, and Epicurean and Stoic philosophers. Luke's picture of Paul's Athenian audience reveals a people less provincial than his previous addressees in Lystra. The Athenians appear enlightened and cultured. Of further interest is the presence of Epicureans and Stoics who seem to hold diametrically opposed views on the role of the Divine in the action of the universe.

Epicurus set up a philosophical school in Athens in 306 BCE. It attracted students to a simple apolitical lifestyle, inclusive of women and slaves. It offered an alternative to Plato's academy, Aristotle's lyceum and Zeno's stoa.[13] Like other philosophers, Epicurus

---

[12] Katherine Bain, *Women's Socioeconomic Status and Religious Leadership in Asia Minor in the First Two Centuries CE* (Minneapolis: Fortress Press, 2014), p. 17, borrowing from the research of Steven Friesen, 'Poverty in Pauline Studies: Beyond the So-called New Consensus,' *JSNT* 26 (2004), pp. 323–61.

[13] See Diskin Clay, 'The Athenian Garden,' in *The Cambridge Companion to Epicureanism*, ed. James Warren (Cambridge: Cambridge University Press, 2009), pp. 9–28.

founded his school in his house, linked to, surprisingly, his suburban garden.[14] This was an ecological space in which Epicureans could come to ponder the philosophical truth about the importance of tranquillity (Gk: *ataraxia*), the heart of the philosophical quest. 'For an unerring understanding of these things [about what is natural and necessary]', writes Epicurus, 'the philosopher knows how to direct every choice and avoidance towards the health of the body and the tranquillity (*ataraxia*) of the soul, since this is the goal of the blessed life.'[15] The inscription at the garden's entrance read, 'Stranger, your time will be pleasant here. Here, the highest good is pleasure.'[16]

The garden offered tranquillity and detachment from the bustle of the urban life that surrounded Epicurus' students. It allowed them to deepen the truth of what they believed. The ongoing value of this ecological space for philosophical reflection is reflected in Epicurus' will. In it he bequeaths the garden to two of his closest friends as an environment for their colleagues and successors to deepen their philosophical learnings: 'I bequeath all my property to Amynomachos and Timokrates, on condition that they shall place the garden and all that pertains to it at the disposal of Hermarchos and his fellow-philosophers and Hermarchos' successors, to spend their time there on philosophy.'[17] Though Epicurus did not believe in a divine being, he placed great importance on friendship as central to coming to tranquillity (*ataraxia*) and the value of visual observation and logic. He was convinced that following ethical principles people could come to live 'a divine life on earth'.[18] The garden was part of this world of ethical discovery.

In this context, then, and with this background, Luke's Paul becomes a philosophical curiosity to the Athenian Epicureans. Their invitation for Paul to say more, especially in the light of the ecological and horticultural context from which their invitation comes, resonates in a way unobserved by commentators on Acts that prepares the ground for Paul's address to them which features a God of creation.

The other philosophical tradition that appears in Luke's Athenian story of Paul is Stoicism. Unlike the Epicureans, the Stoics did not have the luxury of their own place to meet, but were public in their philosophical engagement. In Luke's world, their approach to logical discussion through reason, language and argument, and their questions about the physical world, made them appropriate interlocutors with Paul and the new 'philosophy' that he seemed to be proffering.[19]

Of significance is the Stoic connection between theology and cosmology. Stoics believed that all of creation was intellectually organized by divine reason that permeates the cosmos. From the Stoic viewpoint, 'god is best described as the single active physical

---

[14] For the relationship between Epicurus' house and garden, see Clay, 'Garden,' p. 15.
[15] Epicurus, *Ep. Men.*, 128, as quoted by James Warren, *Epicurus and Democritean Ethics: an Archaeology of Ataraxia* (New York: Cambridge University Press, 2002), p. 3.
[16] Clay, 'Garden,' p. 9.
[17] Richard E. Wycherley, 'The Garden of Epicurus,' *Phoenix* 13 (1959), p. 75.
[18] Margaret R. Hampson, *A Non-Intellectualist Account of Epicurean Emotions* (MPhil, London: University College London, 2013), p. 7.
[19] On the nature of Stoic logic and physics, see John Sellars, *Stoicism* (New York and London: Routledge, 2006), pp. 55–106.

principle that governs the whole cosmos ... [and] the cosmos itself was a living entity and that god was its soul'.[20] To some of the Athenian Stoics who appear in Acts, Luke's Paul appears as a preacher of foreign deities, as they disparagingly ask, 'What is this fellow trying to say (*spermologos*)?' (17.18).

The expression, *spermologos*, levelled at Paul by some of the Athenian intellectuals, judges him as 'a bird-brain devoid of method'![21] Despite this disparagement, Paul is invited to address members of the Areopagus, the meeting place of the Athenian elders (17.16-21), and speak to those assembled, including his Stoic and Epicurean inquisitors. Notwithstanding the critical and arrogant disposition of some, most of his Areopagus audience want to learn from him. 'Can we know what is this new teaching about which you speak? For you bring bewildering things to our ears. Therefore, we wish to know what these things mean.' (17.19b-20). This 'new teaching' and 'what these things mean' concern Earth's Child anticipated a few verses earlier (17.18c). For Luke's audience attuned to Jesus' teaching from the gospel and for contemporary auditors, this will also imply teachings about the use of Earth's goods and the treatment that the Earth receives from its children. More is implied for Luke's audience than a Christological exposé from Paul.

As Paul speaks he draws on themes familiar to his Epicurean and Stoic interlocutors, while managing to present an image of God attentive to the whole of creation and inclusive of all peoples. Paul's speech would appeal most to the Stoics. It becomes one of the great affirmations by the Lukan author of the God of creation. It is an explicit attestation to the Creator God who creates the Earth and permeates all that exists in it, human and non-human. Paul's speech leads to the final proof of the presence of this 'unknown God' through divine patience and forgiveness proved by Jesus' resurrection from the dead. Paul begins acknowledging the deep religiosity of his audience: 'Athenians, I see how deeply religious you are in all things. For as I passed along and noticed your objects of worship I also found an altar with an inscription "To an unknown God". Now the one you reverence as unknown this one I myself proclaim to you.' (Acts 17.22-23).

Paul acknowledges that one of the dedicatory altars that the Athenians worship at is, according to its inscription, 'To an unknown God' (17.23c). This 'unknown God' inscription, like many that dot the major cardo of ancient cities, is carved in stone. The stone, a gift of Earth, provides more than a pedestal for the sacred image. The stone inscription allows Paul to introduce his audience to the One whom Paul worships and they consider 'unknown'. Earth's gift, yet to be fully interpreted through one of Earth's children, will lead Paul to declare the 'unknown God' as the One who is known through creation. Given the ecological richness of Luke's Pauline speech, that flows from the opportunity presented through the stone's inscription, a closer treatment is necessary.

---

[20] Ricardo Salles, 'Introduction: God and Cosmos in Stoicism,' in *God and Cosmos in Stoicism*, ed. Ricardo Salles (Oxford: Oxford University Press, 2009), p. 1.

[21] Pervo, *Acts*, p. 427 quoting Ceslas Spicq, *Theological Lexicon of the New Testament* (Peabody, MA.: Hendrickson, 1994), pp. 268-9.

The Life-Giving God (17.24-28)

The opening proclamation celebrates God's creative action in the cosmos and the effect this has on human beings:

> [24]The God who made the *cosmos* and everything in it, being Lord of Heaven and Earth, does not dwell in temples made by human beings, [25]nor served by human hands as though in need of anything, but gives to all things, life and breath and everything. [26]God created from one human being every human race to live in the presence of everything on the face of the Earth having determined the appointed epochs of time and the boundaries of human habitation [27]to seek God, if only to search and encounter God. And indeed, God exists not far from each one of us. [28]For in God we live and move and have our being, as indeed some of your own poets have said, 'For we also are God's offspring'.
>
> 17.24-28

Paul affirms that God has created everything that exists and is known to humanity (v. 24a, above). He affirms that God is *in* everything. This divine presence permeates the whole of creation, Earth and Heaven. This is the *cosmos* over which God has ultimate authority (v.24b).[22] Luke's attestation about the God of Heaven and Earth echoes the same attribution about God found in the gospel and earlier in Acts, of the communion between Earth and Heaven brought about by Earth's Child, his ministry and ascension, the event that concludes the gospel and opens Acts. The auditor of Acts can no longer reflect on one dimension of the *cosmos* without the other. Heaven and Earth commune with each other as their realities interpenetrate each other through the presence of the risen and now ascended Jesus. He remains with his disciples, Earth's children, as they continue the work of the Spirit through their ministry and, now, through Paul's preaching.

Paul further explicates the implications of God's presence in *everything*. This divine presence brings 'life and breath' to everything (v. 25b). God's breath is *in* everything, a reminder of the Genesis story of creation and the activity of God's breath that becomes the reason for all that exists. If the Genesis creation story lies in the background of Luke's theology of God in Paul's speech, then the ecological implications for contemporary interpreters and auditors to Luke's Acts are clear.

This opening of Paul's speech to his Athenian listeners becomes, in the whole of Acts, a highpoint of ecological recognition. It explicitly articulates the implications of affirming the God of creation whose presence panentheistically infuses everything. As Paul says, God created humanity 'to live in the presence of everything on the face of the Earth' (v. 26). Humans will have this symbiotic relationship with 'everything' that exists

---

[22] On the cosmos, see Hermann Sasse, 'κοσμέω, κόσμος, κόσμιος, κοσμικός,' *TDNT* 3, pp. 867-98, esp. bottom of p. 871, regarding the linguistic association from the time of the pre-Socratic Greek philosopher Anaximander (*c.* 610–*c.* 546 BCE). There are only three other times when *cosmos* appears in Luke-Acts. These are in Lk. 9.25, 11.50 and 12.30. Each instance implies a universal and spatial reality larger than what is confined by geographical boundaries.

on Earth. This means the natural world. God's creative presence has also shaped human existence in such a way that it does not have unbridled power and dominance. Human presence is affected and determined by God's 'epochs of time' and the 'boundaries of human habitation' (v. 26). Further, God's creative and all-pervasive presence in the cosmos is not solely for the benefit of human beings who are limited in time and space. God is not exclusively anthropocentric but embracing of all that exists, including non-human creation. Luke's Paul also recognizes that this God of creation breathes life into everything (v. 25c). Everything, human and non-human, organic and non-organic, is the bearer in some way of God's breath. Thus, everything has the potential to reveal God.

It is possible, in Luke's logic, for human beings who may not explicitly know this God of creation, to come to encounter God precisely through their encounter with creation. Luke is convinced that everything that exists is the fruit of God's breath, God's spirit. To meet creation is to encounter God. As Paul reminds his Athenian audience, 'God exists not far from each one of us' (v. 27c). Luke's Paul affirms, in words drawn from the sixth-century BCE legendary Greek philosopher-poet, Epimenides of Knossos, that this God of creation is also a God of intimacy, and the reason for life, movement and existence, in whom 'we live and move and have our being' (v. 28a). This conviction of God's intimacy, who gives 'life and breath' (v. 25c), leads Luke to conclude this opening section of Paul's Athenian speech with a quote from the Stoic poet, Arastus (315/10–240 BCE), probably well-known to the Acts' audience: 'For we also are God's offspring' (v. 28c). In other words, Luke's view of God, as seen in the gospel, is not one that is purely anthropocentric but totally embracing of everything. Luke's God embraces and breathes life into everything, all of creation, human and non-human, organic and non-organic.

To return to the inscription that launched Paul into this wonderfully rich introduction to the God of creation, the stone inscription is one example of the way that non-organic creation can reflect the power of God's presence. This most inorganic of Earth's elements and the residue of Earth's beginning, speaks. It communicates something of the Athenians' religious desire, of their search of God, and allows Paul to explicate that desire through this meditation on the nature of God. Paul concludes this reflection by drawing out the implications of being God's offspring, the invitation to come to repentance and the assured judgement that awaits, revealed through God's resurrection of Jesus (17.29-31).

On this last point about the resurrection from the dead, there are varying reactions amongst Paul's audience. Some scoff at him; others want to hear more (17.32). The scene ends with Luke identifying a few who become believers, as Paul leaves Athens for Corinth (17.34-18.1).

Conclusion

Paul's speech at the Areopagus is a highpoint of Lukan theology in Acts. It presents Luke's conviction about the Creator God, whose presence can be known through an encounter with Creation. Of all the speeches that Luke formulates in Acts, Paul's presentation to the Athenians is the most explicit ecologically oriented affirmation of the God of creation. The speech anticipates the ultimate reception which Earth's children, and the One about whom they speak, will finally receive at 'Earth's end.' Luke clearly recognizes the opportunity for Earth to provide the means for human beings to encounter God's presence. Within each human being is the desire for God. Luke affirms the Athenians' deep religiosity through Paul's words, after he notices the inscription 'To the unknown God' and their religious objects. He links this to their 'groping' for God in their search for God.

Luke's Paul articulates here the most fundamental of religious and anthropological truths: the desire for religious meaning and communion with God in the various cultural and social settings that confront and confound human beings. Luke believes that people can meet this God, first through 'everything that is on the face of the Earth' (17.26); second, through Jesus. How people will respond to this invitation to divine encounter remains open. Certainly, if the mixed reaction to Paul's speech is anything to go by, then Luke would suggest that the mission 'to the end of the Earth' will not be without its challenges.

Finally, the cultural engagement that Luke presents through Paul at Athens, reflects the attitude which the author has to the Greco-Roman world and the Earth that this world inhabits. The receptivity of Paul's audience to what he has to say that leads to Luke's theological mediation, further mirrors the openness which the evangelist has about the *oikouminē*, its cultures and peoples. The affirmation of this culture comes from the evangelist's own positive disposition to all Earth's creatures, including humans, mediated through the teaching and ministry of Jesus of Nazareth, as revealed in Luke's first volume, and the philosophical traditions represented by Athens' Epicurean and Stoic philosophers and well-known poets.

11

Acts 18.2–20.12. The Artisan, Artemis and the Lord's Supper

The story of Acts that follows on from Peter's Jerusalem defence of his action amongst Cornelius' Gentile householders, that eventually incorporates them into the Jesus movement (15.1-12), and the endorsement that Peter's action receives from the Jerusalem leadership (15.13-29), sets the stage for the expansion of the Jesus movement into the wider Greco-Roman world. As noted in the previous two chapters, this happens under the leadership of Paul as he travels in Asia Minor (16.1-5) and Macedonia (16.6–17.15) and through his engagement with the Athenians in his Areopagus speech (17.16-33). With Paul now centre stage, the figure of Peter no longer appears in Acts.

Paul: Earth's Artisan (18.1-17)

After Paul's ecologically resonant presentation of the Creator God to his Athenian audience, his focus moves on to the Peloponnese Peninsula, to Corinth. Here he connects with two Jewish-Jesus exiles, Aquila and Priscilla, a couple expelled from Rome by the Emperor Claudius (18.1-2). Paul, 'went to see them and because he had the same skill (*homotechnos*), he stayed with them, and they worked together, for as artisans (*technē*) they were tent-makers (*skēnopoios*)' (18.2c-3). This is an important detail that Luke offers us about Paul. The language of artisan or skilled worker applied to Paul and his two companions, Aquila and Priscilla, appears twice (*homotechnos* and *technē*). This is the first (and only time) that Paul's unique tent-making abilities are identified by the author of Acts. Paul never mentions it in his letters. Whether based on Luke's memory of the historical Paul or not, there are several insights that emerge from this skill that add to the ecological import of Paul's ministry in Acts. It expands Luke's portrait of Paul. It confirms in deed what Paul speaks about in his address to the Athenians.

To call Paul a 'tent-maker' (*skēnopoios*) places him amongst those of the artisan class, neither in the elite socio-economic category of the Roman social hierarchy, nor in the lowest, though slaves would have acknowledged his willingness to step down the social ladder by this occupation.[1] Whether or not Paul's occupation represents a

---

[1] Though Joel N. Lohr, ('He Identified with the Lowly and Became a Slave to All: Paul's Tent-making as a Strategy for Mission,' *Currents in Theology and Mission* 34 (2007), pp. 179–87) argues that Paul's tent-making placed him in the social category of slave, and he was thus able to identity with those of the lower and lowest social classes.

Illustration 9 A tent made of goat hair in Wadi Rum, Jordan.[2]

deliberate act of social debasement or renunciation, there are other implications in what Paul did.[3] It helps him to form a sense of solidarity with others who share a similar craft, like the couple Aquila and Priscilla. His tent-making occupation might have been central rather than peripheral to his missionary endeavours, at least as Luke portrays it in this little vignette in 18.2c-3.[4] It gave him a means of financial support, rather than relying on the generosity of other Jesus followers. This craft was also mobile and allowed him to move from city to city while giving him the freedom to preach the gospel. But what was the precise nature of Paul's *technē* for tent-making?

Scholarly opinion is divided as to the nature of Paul's *technē*.[5] Some suggest that Paul was a weaver of *cilicium* (goat's hair), the main material for tents (Illustration 9). Others, that he was a leather worker which allowed him to craft leather products including tents. Perhaps he was both a weaver of goat's hair and a leather worker.[6] A third suggestion comes from considering Paul's 'tent-making' (*skēnopoios*) abilities.

---

[2] Photo by author.
[3] On Paul's 'intense consciousness of debasement' expressed through his tent-making occupation, see Edwin A. Judge 'The Social Identity of the First Christians: a question of Method in Religious History,' *Journal of Religious History* (1980), p. 214.
[4] On the centrality of Paul's tent-making skill to his ministry, see Ronald F. Hock, *The Social Context of Paul's Ministry: Tentmaking and Apostleship* (Philadelphia: Fortress Press, 1980), pp. 66–7.
[5] For this discussion, see Todd D. Still, 'Did Paul Loathe Manual Labour? Revisiting the work of Ronald F. Hock on the Apostle's Tentmaking and Social Class,' *JBL* 125 (2006), pp. 781–95.
[6] Jerome Murphy-O'Connor, *Paul: a Critical Life* (Oxford: Oxford University Press, 1996), pp. 86–9.

This might suggest that, along with Aquila and Priscilla, he constructed temporary awnings, small enclosed huts or open unroofed sheds for short-term events and festivals.[7] But there is a further dimension to Paul's skill that is important to note, and one particularly relevant to this commentary.

Whether leather-worker, goat-hair weaver or constructor of temporary awnings, his skill put him directly in contact with the Earth. The hair of a goat or the skin of an animal came from Earth's creatures that provided Paul with a means to be self-supporting, make connection with other artisans and provide a freedom to preach the gospel. In fact, Luke mentions that 'because Paul was of the same trade [as Aquila and Priscilla] he stayed with them' (18.3). With this seemingly small detail of Paul's occupation, it is clear that the Earth becomes a collaborator with him, freeing him to continue the preaching of the Word which he does immediately amongst the Corinthian Jews and Greeks (18.4).

Paul's preaching is partially successful. It attracts a God-fearer, Titius Justus, living next door to the synagogue, the synagogue head, Crispus, and 'many' Corinthians who are baptized (18.7-8). The reaction from most of Paul's Jewish co-religionists is a different matter. They vehemently oppose his Christo-centric preaching (18.6a). This becomes the moment in Acts when Paul's mission to the Gentiles becomes most clearly stated. In a gesture of denouncement, reminiscent of Ezek. 33.4, Paul expresses Luke's judgement on the Jewish rejection of Paul's message: 'When they opposed and slandered him, he shook out his garments and said to them, "Your blood be on your own heads! I am innocent. From now on I will go to the Gentiles." (18.6-7).

The garment is Earth's symbol that identifies its wearer. The gesture of Paul's shaking of the garment at his recalcitrant Jewish audience becomes an image of release. According to Luke, Paul's listeners bring judgement to themselves in a way that Paul is no longer obliged to them. His obligation to preach to his own people is now freed, given the resistance that they demonstrate.[8] He is released from this obligation, to concentrate on his mission to the Gentiles. However, he resides in a space next door to the synagogue. Whatever of Luke's narrative design to justify Paul's missionary focus on the Gentiles, Paul's presence in this 'next door' space symbolizes the tension that remains in Acts and with its author: the Gentile mission is legitimate and part of the divine plan. But so, too, is God's fidelity to the people of the First Covenant.[9] God is faithful, a theme prominent throughout Luke-Acts. This fidelity includes both Jew and Gentile, as Paul becomes legitimated or designated as God's agent to the Gentiles in a night vision: 'The Lord spoke to Paul at night in a vision, "Do not be afraid, but speak and do not be silent, because I myself will be with you and no one will lay a hand on you to harm you, for there are many who belong to me in this city."' (18.9-10).

It is important to note at this point, as in other parts of Acts that seem to underscore the anti-Jewish tension that surfaces, that this is not literal history, but a theological *apologia* that provides direction for Luke to shape the rest of the story in Acts, especially Paul's Gentile missionary focus. As Johnson notes, 'The formal and artificial character

---

[7] Stanton, 'Accommodation,' pp. 233–4.
[8] Tannehill, *Narrative Unity*, p. 223.
[9] Johnson, *Acts*, p. 323.

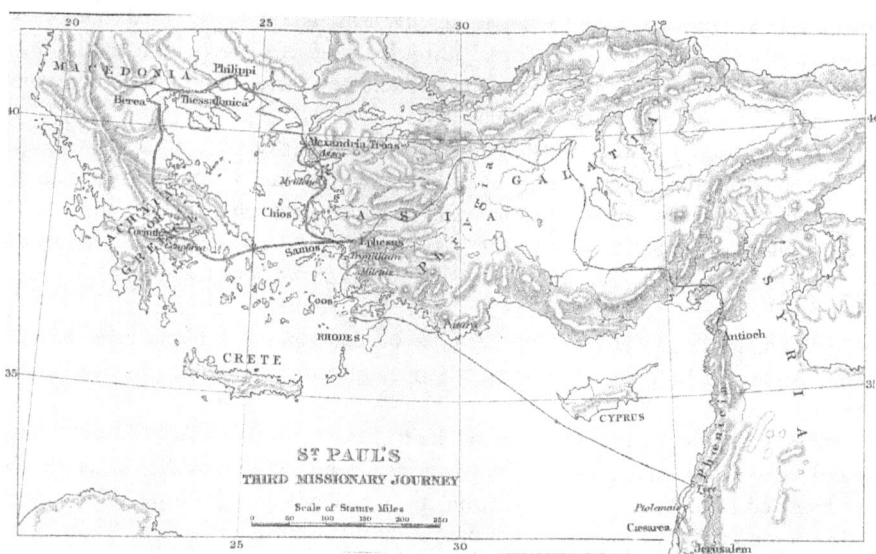

Illustration 10 Paul's third missionary journey.[10]

of these "declarations of turning" by now should be obvious to the [auditor]. They serve to give structure to Luke's story, by making what must have been an extremely complex and never completely resolved tension, into one with a narrative logic and resolution.'[11]

Paul's 'turning' at this narrative juncture in Acts prepares for what immediately follows in the messianic claims and counter-claims put forward before Rome's Corinthian legal representative, Gallio, about Jewish identity (18.12-17). Paul's prolonged eighteen-month stay in Corinth ends not long after the synagogue ruler, Sosthenes, is attacked before the tribunal without any response by the civic authorities (18.17). As Paul concludes his second missionary journey, he travels by ship to Syria, accompanied by Priscilla and Aquila, to Ephesus, where he visits the synagogue and leaves the couple, returning to Antioch via Caesarea, then through the regions of Galatia and Phrygia 'strengthening all the disciples' (18.18-23). As he begins his third missionary journey (Illustration 10), Paul 'passes through the upper region' returning to Ephesus (19.1).

At Ephesus: Artemis and her Silversmiths (19.1–20.1)

The geographical sweep of Paul's journeys contained in these few verses is enormous. Given Luke's brief travelling description, Paul covers almost the whole of Asia Minor, beginning in the east, then north, south and then to Ephesus in the west. This

---

[10] Smith, *Concise Dictionary*, map facing p. 697.
[11] Johnson, *Acts*, p. 326.

Illustration 11 Artemis cult statue, second century CE.[12]

topographical coverage suggests how Earth's terrain, represented so differently in all these parts of Asia Minor, becomes Paul's companion as he moves towards Ephesus, one of the most important cultural and religious centres of the Greco-Roman world. Here one of the seven wonders of the ancient world exists; this is the Artemision, the shrine to the goddess Artemis.

Paul's conflict with those who fostered the Artemis cult, looked at in greater detail below, concludes a complex and important section of Acts set in Ephesus where several important events occur. He formalizes the inclusion of twelve Ephesian disciples into the Jesus movement as he baptizes them 'in the name of the Lord Jesus' (19.1c-7); he again, unsuccessfully, preaches for several months in the synagogue proclaiming the *basileia* of God, but then removes himself and preaches in the hall of Tyrannus, a more public place. This allows 'all the inhabitants of Asia, Jews and Greeks, to hear the word

---

[12] Selçuk Archaeological Museum near Ephesus, Turkey. Photo by author.

of the Lord' (19.8-10). Paul's preaching seems universally effective (to 'all the inhabitants of Asia') and it is accompanied by healing deeds, reminiscent of the actions of Jesus in the gospel and miraculous actions of Peter in earlier chapters in Acts. These confound and defeat itinerant Jewish exorcists and Ephesian magicians (19.11-19). Paul's ministry in Ephesus bears fruit. As a result of all that he says and does, 'the word of the Lord grew and became strong' (19.20).

As Paul contemplates his next journey, to Macedonia, Achaia and finally to Jerusalem, he sends two of his colleagues, Timothy and Erastus, ahead to Macedonia (19.21-22). Luke's focus moves to the economic implications of the Artemis cult. Silversmiths who craft silver shrines of Artemis see their livelihood and the great Artemis temple threatened and disrespected by Paul's teaching (19.23-27). Demetrius, the key instigator of concern for the silversmiths, concludes, 'Not only is there danger that our trade may come into disrepute but also the temple of the great goddess Artemis will be reckoned as nothing and that her greatness will be destroyed, she who is worshipped throughout all Asia and the world (*oikoumenē*).' (19.27). Demetrius' judgement on the implications of Paul's teaching for the prosperity and the worship of Artemis leads to riots, confusion and repeated affirmation of the importance of the cult. The refrain, 'Great is Artemis of the Ephesians', echoes through the whole scene (19.28, 34) as the populace gathers in the theatre. The event and the Artemis-refrain, perhaps capturing something of an historical reality underscoring Paul's Ephesian visit, reflect an attempt to defend a particular socio-economic-political stance, now threatened, and represented through Artemis' worship.[13]

Paul attempted to join the Ephesians in the theatre, and if he succeeded he would not have remained silent. If Paul's speech in Athens is anything to go by, he would argue against the temple and its cult to establish the rightful place of the Creator God. However, Paul's friends restrain him (19.30-31). Eventually, the sane voice of a town clerk quells the crowd and the scene, after two chaotic hours, ends peacefully as Paul encourages the Ephesian Jesus disciples and journeys to Macedonia (19.22–20.1).

Demetrius' conclusion in 19.27 stands for something more than a concern over a loss of income brought about by the teaching of the Way from Paul. It captures a central concern about the role which Artemis plays throughout the whole Greco-Roman world, the *oikoumenē*, as Luke reminds us with an expression discussed in the previous chapter that has ecological implications. As noted there, this *oikoumenē* is more than that part of the universe occupied by human beings. It symbolizes the whole network of interrelated organic and non-organic forms of Earth's existence that affects the lives of human beings.

In this context, Demetrius and his colleagues see Artemis' status as universally significant for the role she plays in the lives of human beings, from which they profit. Though the Artemis cult had local variations, there were common traits to her worship throughout Asia Minor. She was a moon goddess, daughter of Zeus and Leto and twin sister of Apollo. In the Roman configuration, Artemis was Diana depicted with bow

---

[13] Bradley Britner, 'Acclaiming Artemis in Ephesus: Political Theologies in Acts 19,' *The First Urban Christians 3: Ephesus*, eds James R. Harrison and L. L. Welborn (Atlanta: SBL Press, 2018), pp. 127–70.

and arrow, hunting with the wild animals, suggesting a nomadic, pre-agricultural origin.[14] Her association with wild beasts, being called the 'Mistress of Wild Beasts', is one that Luke's Paul will take up later when he addresses the Jesus elders of Ephesus at Miletus (20.29).[15]

In her Ephesian incarnation she is depicted with supernumerary breasts emphasising her as the goddess of fertility, though perpetually virginal, and the Greco-Roman

Illustration 12 Artemis cult statue, first century CE.[16]

---

[14] Tobias Fischer-Hansen and Birte Poulson, *From Artemis to Diana: the Goddess of Man and Beast* (Acta Hyperborea 12. Copenhagen: Museum Tusculanum University of Copenhagen Press, 2009), p. 11.
[15] Homer refers to Artemis as the 'mistress of wild beasts' in *Il.* 21.470. On this 'wild beast'-Artemis' association as a Lukan metaphor for resistance and suffering, which the Jesus followers will experience in the future, to which Paul will address in his speech to the Ephesian elders (Acts 20.29), see Morna Hooker, 'Artemis of Ephesus,' *JTS* 64 (2013), pp. 37–46. Also, Daniel Freyer-Griggs, 'The Beasts at Ephesus and the Cult of Artemis,' *HTR* 106 (2013), pp. 459–77.
[16] Selçuk Archaeological Museum near Ephesus, Turkey. Photo by author.

equivalent to the Anatolian Earth-mother goddess, Astarte or Cybele.[17] Her many breasts symbolized eggs, bull testicles, nuts or signs of the zodiac. Her zodiac neck decoration and animal skirt carvings further emphasized her control over Earth's creatures and power over the cosmos (Illustrations 11 and 12).[18] Within this perception of Artemis' powerful relationship to Earth, its creatures and the whole cosmos, she would be the theological antithesis to the God whom Luke's Paul would celebrate. The restraint which Paul's friends place on him from entering the theatre at Ephesus would have come from a desire not to exasperate the mob-mentality aroused by their dedication to Artemis, rather than from compromising their belief in the presence and power of the Creator God, about which Paul was firmly convinced. This was the God to whom all are attracted and in whom 'all move and have their being' (17.28).

Before concluding Luke's Ephesus episode, there is one final insight that reflects on the ecological implications of the Artemis cult. This links to the auditor's memory of Jesus' gospel teaching about the use of Earth's goods that should generously benefit all.[19] Jesus' teaching about the use of Earth's goods and the way that this is acted out by Jerusalem disciples in the early chapters of Acts, contrast to the way silver, one of Earth's precious metals, is employed to make Artemis shrines for sale to prospective visitors from around the Asian world, where she was most popular. Besides the implication of greedy exploitation understood by Luke's auditors, a further condemnatory echo from Luke's Jesus emerges out of the gospel's narration of Jesus' three-fold temptation and the three ecological principles that emerge (Lk. 4.1-30): the value of caring for Earth's gifts, rather than abusing them through ravaging covetousness; establishing and maintaining a right relationship with the God of Creation through authentic stewardship; respecting Earth's resources without controlling them for selfish gain.[20]

Acts' Ephesian episode with the silversmiths' economic concerns over the potential drop in revenue from the downturn in sales of their Artemis shrines, may seem minor in Luke's overall narrative scheme. The scene taps into larger ecological issues addressed in the gospel. Essentially, at its heart lies the place of God who gives ascetical environmental direction through the teaching of Jesus in the gospel. The predilection of avaricious silversmiths and the elevation of Artemis' status would dethrone the place of Paul's God and Jesus' teaching on possessions. The fact that this tension is not resolved by the time Paul leaves Ephesus never to return means that, for Luke's auditors, this resolution awaits a future when the auditor and contemporary disciple can definitively commit themselves to Jesus' teaching about ascetical discipleship, and centre themselves on the God revealed through the gospel.

---

[17] Johnson, *Acts*, p. 347.
[18] C. L. Brinks, '"Great is Artemis of the Ephesians": Acts 19.23-41 in Light of Goddess Worship in Ephesus,' *CBQ* 71 (2009), pp. 776–94.
[19] For example, in Lk. 4.6-11. See *AEC*, pp. 108–9.
[20] See *AEC*, pp. 107–10.

At Troas: The Lord's Supper (20.2-12)

From Ephesus Paul travels to Macedonia, presumably by ship, and then further into Greece, retracing his steps after learning of a plot against him, and sailing from Philippi to Troas (20.2-6). Here he gathers with other disciples 'on the first day of the week' for the 'breaking of bread' (20.7). This language is evocative of the gospel memory of Jesus' gathering with his disciples for their various meals in the gospel's final chapter (Lk. 24.1, 19, 30, 35), the celebration of the Eucharist and the gathering on the 'Lord's Day' (Acts 2.42, 46; 1 Cor. 16.2; Rev. 1.10), the day when the Jesus movement gathered to remember the resurrection and celebrate Eucharist. More than a commemoration of the Lord's Supper, it is an ecologically redolent celebration of Earth's gifts by which God is celebrated and through which God sustains humanity.

Creation's gifts, the elements celebrated by Jesus with his disciples in the gospels, become the divinely instituted means by which Jesus disciples deepen their communion with the God of Creation. This practice continues in Acts. First noted in 2.42, Paul and his companions continue this Eucharistic practice. It sustains him for his journey and what he is about to face. In fact, it gives Paul so much strength that he is able to preach through the night, at least to midnight (20.7c). But, even for Paul, lengthy preaching has its harmful effects! One of his audience, the young Eutychus, sitting in the window succumbs to a deep sleep and falls three storeys, dead (20.9). Luke describes Paul's response. He goes down stairs to the youth, embraces him, and restores him to life (20.10). Unperturbed by this event, Paul returns to the act of breaking bread and conversation until morning when he leaves Troas (20.11).

Conclusion

The light-hearted and novelistic interlude in Paul's journey in 20.7-12 repeats ecological themes that permeate Luke's gospel: the importance of food, the contribution that Earth's gifts make to sustain the community of Jesus followers, and the act of restoring life and a person to their full humanity.[21] In this scene in Acts, Paul's actions reveal that the *basileia-ecotopia* is present and active.

The scene is not the only one that raises ecological themes in Acts 18-20. The author, almost incidentally, identifies Paul's Earth-related artisan trade (18.2-3). His tent-making abilities, whether from leather or goat hair, puts him in touch with Earth's resources. It also brings him into contact with other similarly occupied Jesus disciples, especially the Roman couple, recently expelled from Rome. The presumption is that this Earth-connection which typifies Paul's occupation and provides for him a modest income, gives him mobility in his preaching to travel to places where his craft would allow him to make contact with people in need. While it seems a parenthetic comment from Luke, his tent-making skills reveal an ecological connectedness and consciousness.

---

[21] See for example, *AEC*, pp. 121, 133, 144, 149, 163, 178, 207-9, 221, 269, 299.

These would accompany him throughout Acts and presumably from the time he first became a committed Jesus follower. Luke never mentions Paul's artistry again. It remains implicit, but an important detail that the auditor remembers.

Luke notes Paul's presence in Ephesus for more than two years (19.10a). This brought him in contact with those associated with the cult of Artemis, regarded as the Earth-Mother, equivalent to Astarte and Cybele of Asia Minor. Time spent in Ephesus would have provided him with the opportunity to preach about Jesus and to critique the cult and those who economically benefited from her worship. In this it seems that he was successful. As Luke notes, 'the word of the Lord grew and became strong' (19.20). The author specifically mentions the silversmiths and the reaction that they, through Demetrius, have to the teaching of the Way explicated by Paul (19.26-27). Though Paul is restrained from entering the theatre, as the Ephesian mob is fired up against those like Paul and his associates, the tension that arises in the story is not just because of the mob violence and the two-hour theatre chant 'Great is Artemis of the Ephesians!'. From the perspective of our ecological hermeneutic, the tension is about the Earth-role associated with Artemis and commandeered by her devotees, especially those who would profit by her cult. Artemis is not the God of Creation exalted by Paul; the ascetical ecological practice of Ephesus' Jesus followers would threaten her silversmiths. For Paul and his friends, Artemis does not exist, and her temple is naught.

Finally, the Ephesian event coupled with Paul's 'breaking bread' and resuscitation of the youth who falls to his death from the upstairs window of the room where Paul has been preaching for hours at Troas, reveal other ecological themes that might go unnoticed. Thus, the ecological teaching and practice of Earth's Child continues in Acts. These will unfold in subtle ways in the final chapters of Acts.

12

Acts 20.13–26.32. Earth's Child Identifies with Earth's Children

After Paul's prolonged time in Ephesus (19.1–20.1) in his third missionary journey and then his short visit to Troas (20.6-12), Luke's key gospel protagonist reveals his intention to travel to Jerusalem in time for the Feast of Pentecost (20.16c). Luke's explicit mention of Pentecost, the most agricultural and ecologically connected feast in the Jewish calendar, reminds the auditor that the creative energies of the Spirit are still present. Paul's desire to arrive in Jerusalem in time for Pentecost also explicitly links his ministry to the Spirit's ongoing empowering and restorative presence first experienced by the Jerusalem Jesus household in Acts 2. This becomes clearer when Paul, en route to Jerusalem, gathers with the Ephesian elders at Miletus (20.17-38).

To the Ephesian Elders: 'Look after Earth's Sheep' (20.17-35)

In Paul's emotive and powerful speech, the evangelist sums up the central elements that have guided Paul's ministry and the imminent concerns which the Ephesian leaders – and all future leaders of the Jesus movement – must expect. The speech becomes a watershed moment. It looks forward and back to what has happened, summarizing for the auditor the central convictions that have shaped Paul's mission. It presents him as an authentic disciple of Jesus, emulating the teachings of Luke's Jesus mirrored in the gospel, especially his spirit of generosity, detachment and ascetical spirituality. This reflects, for the contemporary auditor, an ecologically sensitive character that typifies one who is an authentic member of Earth's children.

Paul declares that his ministerial intention in Asia Minor was to serve God with humility, enduring the sufferings that came to him from his co-religionists and witnessing publicly and privately about God and Jesus (20.18c-21). Paul recognizes that it is the Holy Spirit that guides his future, uncertain of what awaits him as he travels to Jerusalem but firmly convinced that he will continue to witness to the Spirit through the afflictions he will endure as he testifies 'to the gospel of the grace of God', a phrase to which I shall return (20.22-24). Paul acknowledges that this is the last time that he will see his Ephesian colleagues and encourages them to remain as 'overseers' of the 'flock' – Lukan terminology that reflects ecclesial institutional language of a later period – and borrowing imagery from one of Earth's creatures: 'Keep watch over yourselves and over all the flock, of which the Holy Spirit has made you overseers, to shepherd the church of God that he obtained with the blood of his own Son.' (20.28). Paul's phrasing of this encouragement draws from imagery associated with sheep and shepherding, recalling for the Lukan auditor Jesus' gospel teaching about the protection of those who are lost, especially the act of the shepherd caring for the lost sheep (Lk. 15.3-7). This gospel parable, drawn from the familiar pastoral situation of a shepherd looking after sheep, mirrors God's care for humanity and, for Luke's Paul, the image that should govern future pastoral leadership.[1]

This pastoral focus of care sets up the next part of Paul's speech, in which he draws upon an image drawn from another Earth creature representative of the suffering and tension that will surface. He reminds his audience that as leaders they will have to deal with 'fierce wolves' and members of the Jesus movement who will distort truths and confuse others. The metaphor of 'fierce wolves' is an image of the serious tensions and divisions that will come from within and without the Jesus movement. It captures something of their Ephesian experience from those associated with Artemis, the 'Mistress of Wolves'.[2] It is also a reminder of Jesus' gospel teaching to his disciples, of how they will find themselves as 'sheep in the midst of wolves' (Lk. 10.3).

Imagery and metaphors drawn from Earth's creatures provide Luke's Paul with contrasting insights that reveal the future tensions, struggles and turmoil which the Ephesian elders will have to address as they proclaim in their own context the *basileia-ecotopia*. Thus, here in Acts as in the gospel, the author's placement of 'sheep' alongside 'wolves' is a reminder that the Earth community and even members of the Jesus movement, like Earth's non-human creatures, do not live in total harmony but, at times, in angst and tragic division. This insight is a sobering reminder that the *basileia-ecotopia* has not arrived.[3] Its completion awaits a future for which the Paul of Acts labours.

Paul the Ascetic (20.33-35)

As Paul moves towards the final parts of his speech, he commends his listeners to God and to the 'grace of God's word' (20.32b). Gospel auditors would know that this 'grace', to which Paul had referred earlier in his address (20.24c), is not a commodity that one earns, but an encounter with the very being of God that brings deep communion. This communion creates unity amongst human beings and their relationship to Earth and its creatures. Grace is of the essence of the *basileia-ecotopia*. In the rabbinic literature it reveals God's glory.[4]

Within this context, Paul (or, rather, Luke) affirms the ascetical, generous and Earth-affirming spirit that has guided his gospel labours and addresses the imbalance of wealth that he would come across as a tent-maker in his journeys:

---

[1] See *AEC*, p. 211.
[2] Hooker, 'Artemis,' pp. 37–46.
[3] See also *AEC*, pp. 170-1.
[4] Gerhard Delling, 'πλεροω,' *TDNT* 6, pp. 286–311.

³³I coveted no one's silver or gold or clothing. ³⁴You know for yourselves that I worked with my own hands to support myself and my companions. ³⁵In all this I have given you an example that by such work we must support the weak, remembering the words of the Lord Jesus, for he himself said, 'It is more blessed to give than to receive'.

20.33-35

With this final word, he kneels and prays with his companions for a last time. They kiss and embrace him, illustrating the very essence of the communion brought about by the grace of God, illustrative of the *basileia-ecotopia*. They finally escort him to his ship and Luke moves Paul to the next part of his travels and the completion of his third missionary journey (20.36-38).

The Voyage to Caesarea (20.36–21.17)

Returning to the Pentecost-theme and Luke's overarching narrative intention in this section of Acts, the auditor remembers how Paul desires to be in Jerusalem for its celebration. It is noteworthy how in just a few verses Luke narrates Paul's ship journeys. He has travelled from Troas and Assos to Miylene, Chios, Samos, sailing past Ephesus, arriving at Miletus where he addresses the Ephesian elders (20.13-16). Now he sails from Miletus to Cos, Rhodes, Patara and Myra, and then on to Phoenicia, sighting Cyprus, finally arriving at Tyre. This is an extensive sea voyage which Paul undertakes, along the west coast of Asia Minor and across the Mediterranean towards Phoenicia. It again evokes, for the ecologically sensitive auditor, all the environmental connections about ships and sea travel raised in the introduction.

The short phrase that Luke appends to the ship's arrival at Tyre, 'for there the ship was to unload its cargo' (21.3c), is a further reminder of the function of cargo shipping in the ancient world, the way commodities gleaned from Earth's resources and distributed around the Empire through its shipping network, offered the opportunity for Earth's gifts to be shared with those in need and, at the same time, enabled merchants and the elite to benefit financially. The unloading of cargo at Tyre's port duplicated what happened in all ports around the Mediterranean controlled by Rome. It helped reinforce the socio-political hierarchy of Roman social stratification.[5]

After a brief interlude with the disciples of Tyre, with a scene of farewell similar to that which Paul experienced from the Ephesian elders at Miletus (21.5-6), Paul with his companions (the 'we') continue their sea journey to Ptolemais, arriving finally at Caesarea (21.7-8). In a brief sojourn at Caesarea, Paul stays at the house of one of the seven, Philip, now called 'an evangelist' (21.8) who is noted as having four unmarried daughters who prophesy (21.9). While at Caesarea, a Judean prophet, Agabus, warns Paul in a prophetic symbolic action, binding his hands and feet with his own belt, about the impending suffering he will undergo in Jerusalem at the hands of his co-religionists

---

[5] See *AEC*, pp. 26-30.

(21.10-12). But Paul's commitment to go to Jerusalem remains steadfast with words that echo the gospel's Lord's Prayer (Lk. 11.2-4), 'The will of the Lord be done' (21.13).[6] With this resolve Paul leaves Caesarea accompanied by Jesus followers who accompany 'us' to Jerusalem (21.15-17).

In a meeting, James and the Jerusalem elders summarize the criticism levelled at Paul by some of his Jewish compatriots, who believe that he has compromised his Jewish heritage and Torah observance (21.18-21). The Jerusalem leaders recommend Paul undertake a public act of purification to prove his Jewish commitment (21.22-24).[7] However, even this does not prevent some antagonists from falsely accusing him of contaminating the temple (21.27-29). In an episode reminiscent of Jesus' passion (Lk. 22.47–23.25), they seize Paul, drag him from the temple but, before they can stone him, the tribune's cohort arrest Paul, bind him and bring him into the military barracks away from the violent mob, where he is falsely accused of Messianic pretence (21.31-38). With the permission of the tribune, Paul addresses the crowd (21.39-40).

Paul's Jerusalem Address: About the Ascended Jesus (22.1-21)

Paul's address in Hebrew, his *apologia*, quietens the people. He first recalls his Jewishness, the place of his birth in Cilicia in Tarsus, his religious education by the rabbinic teacher, Gamaliel, and his persecution of members of the Jesus movement ('the Way') with the encouragement and witness of the Jewish religious leadership (22.3-5).[8] Paul begins to recall again (see 9.1-29) the moment in which he meets the risen Jesus, an encounter that permanently alters his life's direction.

The images that Luke uses are ecologically rich. Paul is on a journey – an explicit image of discipleship that recurs through the gospel and links to the metaphor of the 'way' – towards Damascus. Heavenly light, a 'great light' (22.6), shines on him, he falls to the ground, and a voice from Heaven asks Paul the reason for his persecution of the one who speaks (22.7). Paul recognizes the heavenly status of the voice as he asks, 'Who are you, Lord?' 'I am Jesus of Nazareth whom you persecute' (22.8). The heavenly voice is the exalted and ascended Earth's Child who identifies with those members of the 'Way' whom Paul persecutes. The auditor remembers how Jesus' presence and ministry in Luke brought together the conventional spheres of cosmic identity, Earth and Heaven. These co-penetrate and merge, an insight discussed previously, especially in Chapter 1. In Acts, this communion of Heaven and Earth continues. In the beginning of Acts, Earth's Child ascends to be with God of the Heavens. His presence continues

---

[6] For the ecological implications of the Lord's Prayer, see *AEC*, pp. 178–80.
[7] The Jerusalem Jesus leaders also indicate how they have responded by letter to the Gentile seeking membership into the Jesus movement, originally Jewish (21.25). The letter is a repeat of the action already undertaken earlier in Acts (15.20), suggesting that '[i]t is possible that Luke is, in one direction or another, altering the historical reality, or, more likely, exercising his accustomed authorial right to remind the [auditor] of important points': Johnson, *Acts*, p. 376.
[8] Besides this time in Acts when this expression for the Jesus Messianic movement, 'the Way', occurs, the author also uses it at 9.2, 19.9 and 19.23.

through the Spirit's empowering action which accompanies the members of the Jesus movement as they travel beyond Jerusalem into the Greco-Roman world.

The Spirit accompanies Paul in his ministry and travels. As Paul addresses his Jerusalem audience in Acts 22, and looks back on what has happened, he recounts how the One from Heaven, born of Earth, speaks from Heaven. This causes Paul to fall to Earth (to the 'ground': 22.7). This 'forced' communion with Earth's elements by the action of the heavenly voice opens Paul up to the divine encounter. In his experience of Heaven's voice and Earth's ground, these two spheres, now one, become the means by which Paul comes to a new awareness of his relationship with Earth's Child and with those whom he persecutes. Ecological encounter brings Paul to a conversion of heart and a new vision. The identity, intimate communion, which the risen and ascended Jesus, Earth's Child, has with those whom Paul persecutes makes them Earth's children. These are the ones whom Earth's Child now protects.

Paul recounts how those who were with him on the road to Damascus also see the light, though hear no voice (22.9). Paul seeks direction from the heavenly One (22.10) who tells him to go to Damascus, and meet a devout Jew, Ananias, from whom Paul receives sight after becoming blind from the light (22.11-13). Ananias, Paul recounts, affirms him as God's witness and baptizes him, formally bringing him into the Jesus movement and membership of Earth's children (22.14-16). Paul then stresses how this moment of divine encounter confirmed in baptism sets the stage for his return to Jerusalem, his rejection there and a divine commission to go to the Gentiles (22.17-21).

Earth and Existential Life (22.22-23)

Up to this point in Paul's speech, his audience is silent. But his affirmation of his divine commission to the Gentiles leads to a violent threat from his listeners: 'Away from the Earth (*gē*) with this fellow, for he should not be allowed to live!' (22.22). There is a critical truth expressed here which Paul's audience offer, however unwittingly. They affirm that one's connection to Earth (*gē*) is a source of life. Their instinct is to remove this character, who seems so connected to Earth and its gifts that come from his ministry on behalf of Earth's Child, from communion with Earth. Ecological separation from what nourishes and sustains through Earth can lead to existential death. This is an important insight that Paul's antagonists state. Earth-connected gestures, waving their garments and throwing dust into the air, testify to their rage (22.23).

The tribune again rescues Paul and decides to interrogate him privately in the barracks, away from the crowd, with torture (22.24). It is at this moment that the auditor and Paul's torturer learn that Luke's Paul is a Roman citizen, underserving of such treatment (22.25-29). On the following day the tribune establishes a council of the Jewish religious leadership in an attempt to establish the reasons for Paul's accusation (22.30). Paul's speech to the council again reaffirms his divinely endorsed credentials, argues against the authenticity of the Jewish leadership, and ends up creating a violent riot within the council because of the different theological perspectives held by the Pharisees and Sadducees (23.1-9). Again, the tribune extracts Paul from the violence that surrounds him, and brings him to the barracks (23.10).

Illustration 13 The remains of the ancient harbour of Caesarea Maritime. Paul leaves from here for his final voyage to Rome.[9]

That night God speaks to Paul, encourages him in his witnesses and attests that he will bear witness in Rome (23.11). The divine plan is laid out. This will guide Paul's journey to the end of Acts when he finally arrives in Rome. Despite a plot from his co-religionists to kill Paul in an ambush (23.12-15), the attempt is thwarted with the help of his nephew and the tribune's rearrangement of Paul's escort to Caesarea and its governor, Felix (23.16-24). Paul lingers in Caesarea's prison for two years after a series of trials before the Roman authority from various representatives of the Jewish Jerusalem leadership (23.25–25.8). Finally, when Paul is asked by the new governor, Festus, if he would go to Jerusalem to face his accusers and defend himself, Paul attests to his innocence and his conviction of the falsehood of the charges against him (25.9-11). He appeals to be tried as a Roman citizen before the Emperor Caesar. Festus grants his wish (25.12).

Paul's Final Trial in Caesarea (25.13–26.33)

One last trial, attestation and *apologia* await Paul before he can undertake his final voyage from Caesarea (Illustration 13) to Rome (25.13–26.32). This happens in the presence of King Agrippa and his wife Bernice who have also come to Caesarea (25.13).

---

[9] Photo by the author.

Luke's presentation of the events that unfold before Agrippa attest, like Jesus in his own gospel trial, Paul's innocence, his Jewish heritage and commitment, his persecution of Jesus followers and, again with ecologically infused imagery, his encounter with the heavenly divine light, which caused him, and all those accompanying him to Damascus, to fall to Earth ('the ground') (25.14-18). Luke again summarizes Paul's experience, of the vision and its accompanying voice, that becomes the source of his commission to turn to the Gentiles. Paul then explicates the implications of this sacred encounter, of its faithful reflection that has its origins back with Moses and the prophets, and comes to its final expression in the death and resurrection of Jesus (26.19-27). Despite the perception that Paul wants to make Agrippa a 'Christian' (26.28-29), the final Roman decision, represented in Agrippa, Bernice and Festus, is reached: Paul is innocent and could have been freed had he not appealed to Caesar (26.30-32). To Rome he will go.

Conclusion

This is a very ecologically rich section of Acts, as Luke moves Paul from Greece, Asia Minor, back to Jerusalem, and the end quest of Paul's hopes to be in Jerusalem to celebrate Pentecost. Behind Paul's intention is Luke's purpose to affirm the work of the empowering Spirit that has, hitherto, confirmed the spread of the Jesus movement outside Jerusalem and beyond Judea. Paul has been the Spirit's principal protagonist. The reminder about the importance of Pentecost returns in this section of Acts. Given all that has happened in this 'in-between-time' through Paul, between the first account of Pentecost in Acts 2 and its reminder here, there is the intention to return to the beginning, the origins of the movement of Earth's children initiated by the descent of the Spirit that occurred in the first Pentecost.

To the contemporary auditor attuned to the ecological implications of Luke's Paul, three points can be noted. First, over these chapters, Paul's ascetical practice and generosity are highlighted. They reveal one steeped in the teaching of Luke's Earth Child. His address to the Ephesian elders affirms his spirit of material detachment that allows him to minister in a way that is free and not driven by any intention other than to serve God with humility (20.19), a profoundly deep ecological characteristic of a disciple imbued with Jesus' teaching about possessions. This spirit guides Paul's explicit ministry to the poor and the weak (20.35).

Second, in this section of Acts and in Acts 27, our focus in the next chapter, Paul and, at times, his companions (the 'we' or 'us') undertake major ship voyages across the Aegean and Mediterranean seas. The ships involved in these voyages with their cargo are of important ecological significance in the Greco-Roman world. I shall not go into them in detail here. They will be part of my focus in the next chapter and the final two chapters of Acts, where Luke has Paul undertake a final, dramatic and adventurous great sea voyage that will bring him to Italy and the road that leads him to Rome. The little note, almost by way of passing, which Luke makes of the ship berthed at Tyre's harbour, 'for there the ship was to unload its cargo' (21.3c), is a brief reminder of the way and need that Earth's resources become transported around the Roman Empire.

Third, the section of Acts that is our focus in this chapter takes up the various accusations levelled at Paul and the *apologia* that Luke's Paul provides in his defence. Twice Paul narrates his account of his encounter with heavenly light and the voice that accompanies it (11.5; 26.12-18). Though both accounts have slight variations in detail, they attest to the role which Earth and Heaven, the two once-separated now-joined spheres of cosmic existence, play in Paul's change of direction. He moves from being a persecutor of the Jesus movement to becoming one of its most ardent members as he learns that Jesus, Earth's Child, identifies with those who belong to him, Earth's children. Earth's Child feels for the suffering of Earth's children. Their sufferings are his.

Part Three

13

Acts 27.1–28.31. The Final Voyage towards Rome and Earth's 'End'

In these last two remaining chapters of Acts, Paul undertakes the most daring of voyages (Illustration 14). Luke's description of these is filled with dramatic detail. The discussion on ships and challenges of sea travel around the Mediterranean from Chapter 8 is a most pertinent background for what happens to Paul as he journeys to Rome. The agenda which drives this closing section of Luke's absorbing narrative concerns the determination by the evangelist's hero to get to his divinely predicted destination. Paul's resolve, despite all the difficulties he encounters in his journey, attests to his fidelity to God and Jesus to complete his mission. Applying an ecological hermeneutic to these final chapters in Acts allows the role which creation and Earth's elements play in bringing Paul to Rome to emerge. Luke Timothy Johnson notes, 'Luke's narrative as a whole is an *apologia* for God's work in history, and that theme is certainly present in this account'.[1] Luke's voyage account of Paul's final journey to Rome is also an *apologia* for God's work which involves creation and Earth.

Rome, Paul's destination, is not only the centre of the Roman Empire, it is also an ecological symbol. While many studies concern Rome's social, architectural, and cultural construction, Rome is symbolic of the environmental control which its political leaders, especially the Caesars, had over the Empire. The scene that concludes Acts, and the evangelist's two-volume oeuvre, has Paul under house-arrest in a domestic setting. This is a subversive ecological image in a world subservient to the Empire's urban centre and principal city. It is this concluding image which offers hope for Earth's children and the rescue of 'Earth's end', a point to which I shall return towards the end of this chapter.

The Initial Voyage and Earth's Reaction (27.1-8)

The earlier appeal that Paul makes to Festus to have his case against him for political and religious sedition tried before Caesar in Rome (25.11-12), and later confirmed by Agrippa, now becomes realized. Paul, his companions (the 'we'), and 'other prisoners' are placed under the charge of a high-ranking centurion, Julius, who will escort them

---

[1] Johnson, *Acts*, p. 458.

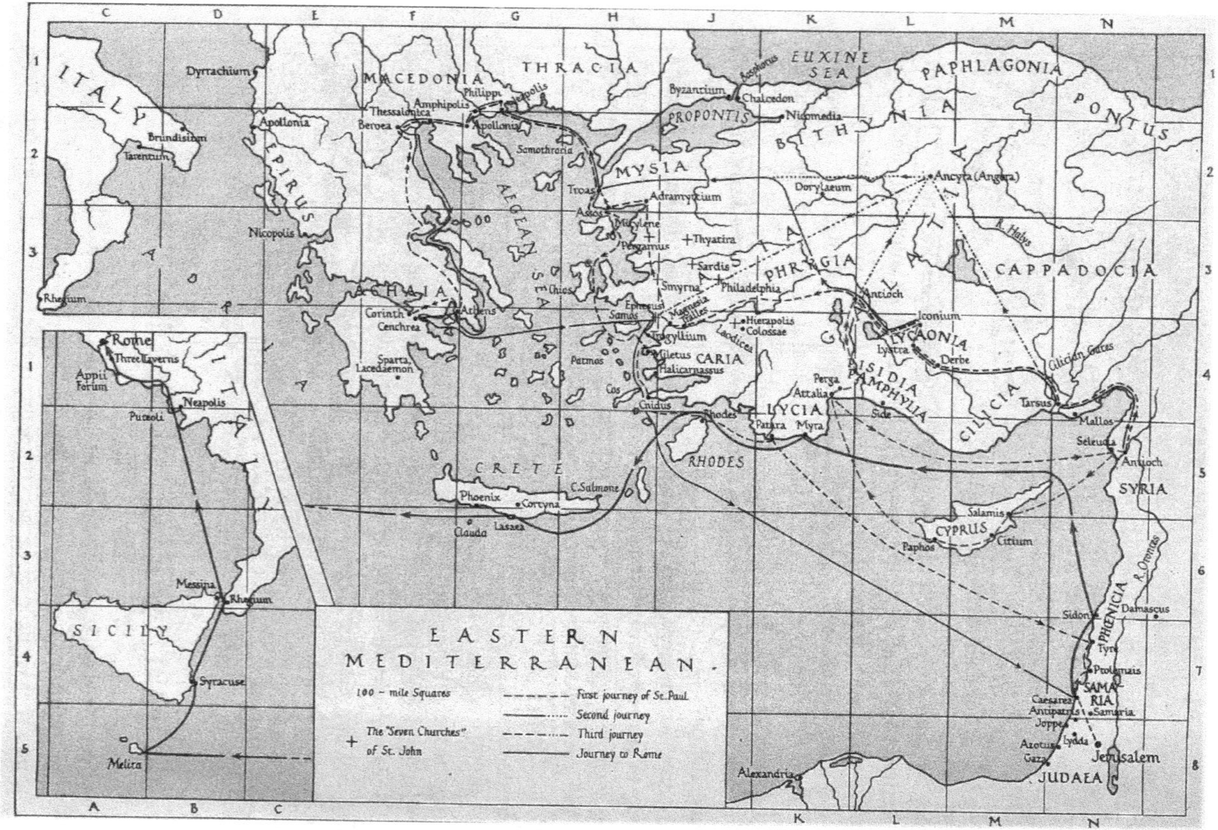

Illustration 14 Paul's sea journeys including his final voyage to Rome.[2]

[2] Map from Bernard Orchard, Edmund F. Sutcliffe, Reginald C. Fuller, Ralph Russell (eds.), *A Catholic Commentary of Holy Scripture* (London: Thomas Nelson and Sons Ltd, 1953), p. 1312.

on their voyage to Italy (27.1). They first board a coastal vessel which takes them to Lycian Myria via Sidon, the lee of Cyprus, that offers protection from contrary winds, and the waters off the coast of Cilicia and Pamphylia (Illustration 14; 27.2-5).

Luke's brief description of this initial journey, with the mention of the difficulty the ship has in sailing against adverse winds, anticipates the travelling challenges and potential dangers that will unfold. More than that, these anticipate the parallels that Luke will draw between Paul's voyage and the experience which Jesus' disciples have in the gospel as they sail on the Sea of Galilee (Lk. 8.22-25). Ecological resonances are present in the way that Earth's seas, affected by the winds, have a power and presence that can assist or restrain human intention. As Luke's voyage narrative gathers momentum, Earth's elements seem to threaten to annihilate Paul and his travelling companions, a similar threat faced by Jesus' disciples in their Galilean sea crossing. What will be essential for Paul is the manner of a divine exorcism that will calm Earth's waters and allow for Paul's eventual safe passage to Italy.

At Myra, Julius moves his charges on to a larger Alexandrian grain-carrying vessel bound for Italy (27.6).[3] The ecological implications of this nature of vessel and its cargo, human and agricultural, are clear. Its passengers and grain cargo are dependent on the wooden structure that carries both across the seas and over waters potentially fraught with danger. They are subject to Earth's weather that accompanies the voyage. The fragility of Earth's creatures who venture on to the Mediterranean Sea is palpable. This is confirmed in the manner by which the ship sails slowly, arrives with difficulty off Cnidus, and experiences contrary winds that prevent it from sailing by the usual navigational route. It alters its path, sailing under the lee of Crete, off Salmone, before coasting to the 'Far Havens' harbour close to Lasea (Illustration 14; 27.7-8).

The 'Dangerous' Voyage (27.9-20)

At this point, Luke mentions the loss of time because of the unfavourable sailing conditions experienced. But there is another description which the author gives the voyage. It is 'dangerous' especially given the weather conditions at the particular time of year that they sail: after the 'fast', after the Day of Atonement, which took place around the autumnal equinox (27.9).[4] Paul gives voice to the author's judgement preparing the auditor on how the narrative will unfold: 'Men, I see that the voyage will be dangerous and much loss, not only of the cargo and the ship, but also of our lives.' (27.10).

Paul's concern is holistic. He is concerned, of course, that the voyage that they have undertaken will be dangerous for all its passengers. They could die. His concern is also for the ship and its cargo, products of Earth and gifts for the welfare of the Roman people at whose port near Rome the ship will berth with a cargo that offers sustenance for the well-being of its people. The ship's potential destruction has ecological implications. Paul's holistic ecological concern, however, contrasts with the lack of

---

[3] Winter, 'Acts,' pp. 60-1.
[4] Murphy-O'Connor, 'Travelling Conditions,' p. 45.

interest which the centurion has for the ship, its cargo and passengers. His focus is elsewhere, pragmatic, and more anthropocentric. He is more interested in the ship's captain and owner, to set sail hastily, rather than a concern for the safety of the ship, its occupants and cargo (27.11).

For this reason, the ship continues its voyage despite the danger acknowledged by Luke and confirmed by Paul. It moves from the harbour in which they have berthed, unsuitable for winter, to Phoenix, a harbour in Crete that promises better protection (27.12). A favourable wind initially allows the ship to sail close to Crete's shoreline and aim for Phoenix (27.13). But a 'tempestuous' wind from land later blows the ship towards Cauda, a small island (27.14), as its passengers begin to feel more concerned about their precarious situation (27.15-17).

The story now unfolds in two parts. The first describes the travellers' response to the violent storm, its effect on the vessel, the encouragement which Paul offers his travelling companions, a meal he shares with them, and their attempt to rescue their ship (27.18-38). The second describes the inevitability of the ship's destruction as it runs aground and what happens to its human and grain cargo (27.39-44). Luke begins the first part of the story by intensifying the drama of the moment: 'As the storm violently pounded us, on the next day they began to throw the cargo overboard, and on the third day they threw the tackle overboard with their own hands. When neither sun nor stars appeared for many days and no small tempest raged all hope of our being saved (*sozō*) was finally abandoned' (27.18-20).

The evangelist describes the hopelessness of the situation in terms of salvation. The verb that Luke uses for 'being saved' (*sozō*) implies more than a loss of life. It reflects the deep existential quest for communion with God and the desire to be liberated from a situation that threatens to annihilate them. It appears that this deeper communion with God and the release from the impending disaster seems compromised. Human desperation parallels the cosmic upheaval of the scene as sun and stars disappear, and darkness dominates over raging waters that echo the cataclysm. Significant is the synchronicity between humanity and creation. It would seem that the peaceful coexistence between Heaven and Earth, especially its creatures, is seriously disturbed. The scene illustrates the desperation.

Paul's Vision and Shared Meal (27.21-38)

In this context, Paul offers an alternative vision that rescues the lost hope that pervades the scene's characters. He castigates them for not listening to him (27.21-22), encourages them to take heart, predicts that there will be no loss of life, only the ship, and shares with them God's nocturnal communication to him in the form of an angelic vision that assures him and his companions that they will safely complete their voyage (27.21-25). But first, says Paul, the ship will have to run aground (27.26). The drama that unfolds next affirms Paul's prediction. In the middle of the night, the ship inches towards land, confirmed by depth soundings, with its possibility of being wrecked on rocks (27.27-29). The sailors look for a way to escape the ship and the impending calamity. Paul reminds the centurion and his soldiers, 'Unless these remain in the boat, you cannot be saved

A	Paul's sailing companions without food (27.21)
	The events that surround Paul, the ship and his sailing companions (27.22-32)
A¹	Paul's sailing companions with Eucharistic food that strengthens (27.33-38)

Figure 11 The theme of food that frames the ship saga (Acts 27.21-38).

(*sōzō*)' (27.31). In other words, within the Lukan salvation schema, which the author establishes throughout the two-volumes – first in the gospel story of Earth's Child and now, in the extension of this story in Acts concerned with Earth's children – salvation is not a private, individualized prerogative. Paul's injunction to 'remain in the boat' theologically affirms that salvation – the rescue from all that threatens to annihilate – comes through solidarity with the human community and Earth's gifts, represented by a floundering ship, no matter how fragile these may seem to be. As Luke's Jesus teaches, salvation also emerges for potential disciples in their ability to be released from dependency of material possessions. The soldiers' act that immediately follows, of cutting the ropes that secure the boat and letting it go (27.32), symbolizes this teaching.

Further, the theme of food (27.31, 33-38) frames this whole episode in Acts 27.21-38 (Figure 11). Before Paul offers his insights and encouragement from his nocturnal vision (27.21-26), Luke notes how the travellers had been a long time without food (A-27.21a). At the episode's conclusion, after the soldiers sever the ropes securing the boat, Luke mentions the arrival of dawn and Paul's encouragement to his companions to take food (A¹-27.33-38). The kind of food that Luke thinks of here is not ordinary food. Paul's feeding of his hungry companions in a desperate situation seems Eucharistic; it parallels Jesus' miraculous feeding of the hungry crowd in Lk. 9.10-16.

> ³³Just before daybreak, Paul urged all of them to take some food (*trophē*), saying, 'Today is the fourteenth day that you have been in suspense and remaining without eating having taken nothing. ³⁴Therefore I urge you to take food (*trophē*), for (*gar*) it is for your salvation (*sōteria*); for (*gar*) none of you will lose a hair from your heads.' ³⁵Having said this and taking bread, he gave thanks to God before them all, and, breaking it, he began to eat. ³⁶Then all of them being encouraged also took food (*trophē*) for themselves. ³⁷We were in all two hundred seventy-six souls in the ship. ³⁸Being satisfied by the food (*trophē*), they lightened the ship by casting the wheat into the sea.
>
> 27.33-38

There are several important theological features to this scene: (a) Paul urging his friends to eat; (b) the salvific and protective importance which Paul attaches to the act of eating; (c) the repetition of the word for 'food' (*trophē*) throughout the incident; (d) the manner in which Paul eats; (e) his companions' participation in the meal; (f) their satiation and restored strength from eating that helps them to lighten the weight of the ship; (g) finally, they 'cast' the grain cargo into the sea.

(a) Paul exercises a style of leadership amid this dire situation. The particular aspect of that leadership in this moment centres on Paul's encouragement to those who are despondent and desperate, fearing for their lives. Paul's encouragement gives his sailing companions heart. He shows concern for their physical safety and spiritual well-being. This link, the connection between the physical and spiritual, undergirds the importance and value of creation and Earth. In the narrative, 'food' becomes the symbolic signifier of this connection which holds deep theological meaning.

(b) Luke explicates the theological significance of food-sharing in v. 34. Two consequences come from this act. Both are indicated in Paul's words and Luke's structure of the sentence with the repetition of the word 'for' (*gar*):

> I urge you to take food,
> *for* (*gar*) it is for your salvation (*sōteria*);
> *for* (*gar*) none of you will lose a hair from your heads.
>
> 27.34

First, Luke's Paul links food to salvation (*sōteria*). Some translations consider *sōteria* to mean 'safety' or 'survival'.[5] But it is more than that. *Sōteria* ('salvation') is an important theme in the Lukan corpus which Lukan scholars consider central to Luke's project mediated through Earth's Child and continued in the ministry of Earth's children.[6] The word occurs frequently in Luke-Acts.[7] It concerns the redemption and liberation of all Earth's creatures, including human beings. This is certainly explicit in Paul's encouragement to his companions to eat.

Second, their act of eating brings about another consequence: they will know God's protection, as Luke's Paul draws on teachings from Jesus familiar from the gospel. Jesus taught his disciples that they can trust the sustaining and caring presence of the Creator God, who cares for all creation and will protect God's beloved ones, despite the presence of even physical death (Lk. 12.22-34; 21.18).[8] The implication of Jesus' teaching is palpable in the situation in which Paul repeats Jesus' words: his audience is facing certain death, but with the food that Paul offers they will know both *sōteria* and God's protection that will not allow them to 'lose a hair' from their heads.

Food, formed from the fruits of the Earth, offers the boat's passengers sustenance and a definitive bond with Earth's God. With this deeper soteriological connection, food becomes a necessary means to communion with and protection by God. It will strengthen them to respond to the destructive forces caused by the wind that, like the exorcism enacted by Jesus against the storm on the lake (Lk. 8.22-25), needs exorcising.[9]

---

[5] For example, the translators of NRSV take the first part of this sentence to mean 'I urge you to take some food, for it will help you *survive*'.
[6] For example, Jacques Dupont, *The Salvation of the Gentiles* (New York: Paulist Press, 1967), pp. 11–34; I. H. Marshall, *Luke: Historian and Theologian* (Exeter: Paternoster, 1970), pp. 77–215.
[7] Lk. 1.69-71, 77; 2.11, 30; 3.6; 19.9-10; Acts 2.21, 40, 47; 4.12; 5.31; 11.14; 13.23, 26, 47; 16.17, 30-31; 28.28.
[8] See *AEC*, p. 257.
[9] See *AEC*, p. 154.

(c) As Paul urges his companions to eat, Luke refers to 'food' (*trophē*) four times (27.33, 34, 36, 38). This repetition is not accidental. Food and the meal in which food is taken is a communal rather than private act. It is the sharing of Earth's gifts and ultimately the fruit of God's action. The communal setting, which Luke symbolizes in the image of a boat,[10] also reveals God's kindness on humanity through the sharing of Earth's gifts. This communal action comforts and strengthens those who partake. Creation (which the auditor hears in the repetition of 'food') is an actor in the event. This connection, of Earth's gifts (*trophē*) being an active agent in the liberation of humanity in a communal setting, is explicit in Jesus' meal ministry in the gospel.[11] This is one of Jesus' principal acts of ministry that reveals the kind of God whom Jesus celebrates and proclaims. Luke's four-fold mention of 'food' and the act of communal eating have a deeper function that is more than simply pragmatic, to satisfy hunger pangs. It is also *theological*, as seen above.

(d) The manner of Paul's eating reminds the auditor of the way Jesus provides food for the hungry crowd (Lk. 9.10-17), eats with his disciples in their final meal (Lk. 22.28-34) and in the post-Easter meal at Emmaus with two perplexed disciples (Lk. 24.28-29, 32). In these three gospel stories, the meal surrounds words and deeds of encouragement, the theme which Luke explicates in Paul's meal with his desolate companions. Similarly, Paul's actions – 'taking', 'giving thanks', 'breaking' and 'eating' – repeat Jesus' actions in these stories. In all these meal-stories in the gospel and in Acts 27.33-38, Jesus and Paul take 'bread'.

(e) The combination of all these features in the story suggests that Paul's meal in front of his companions is more than a simple Jewish meal, as some have suggested.[12] It is clearly Eucharistic.[13] Paul's meal companions do not just repeat what Paul does, only eating *after* Paul has set an example, which some translations suggest ('Then all of them were encouraged and took food for themselves': 27.36, NRSV). Rather, Paul encourages them to participate in the same meal which he inaugurates. They become, literally, *companions* with him. They eat the blessed 'bread' *with* him.

(f) Verse 37 ('We were in all two hundred seventy-six souls in the ship') further underscores the communal setting of the meal and the satisfaction the food brings (27.38a). Luke's precision about the number of passengers in the boat who partake of the meal is intriguing. Is it a precise recall in early morning light knowing that the number on board the vessel would be double that number? Does the author suggest by the number and the implications of the lack of a full passenger-list, the level of tragedy and drownings that had occurred prior to Paul's encouraging words and actions?[14] Or do the numbers suggest those who participated in the meal were relatively larger than the number that Luke's audience would know would usually participate in a household arrangement for the Eucharistic meal?[15] Or is Luke reinforcing the spirit of the

---

[10] Lk. 8.22-25. See also *AEC*, p. 154.
[11] *AEC*, p. 20.
[12] For example, Dunn, *Acts*, p. 341.
[13] See Johnson, *Acts*, p. 455, footnote 35.
[14] Dunn, *Acts*, p. 341.
[15] See *AEC*, pp. 35, 160.

communal act of 'breaking bread', which typified the early Jerusalem community of Jesus followers who participated in the 'breaking of bread' (2.46). Or all of these? Whatever the exact meaning, it is clear that in this moment of crisis the Eucharistic meal becomes the centrepiece for those who share it, to strengthen them, give them courage and enable them to take the next step in overcoming their present disillusionment, and lighten the ship. This same pattern (disillusionment – Eucharistic meal – clarification – action) occurs in Luke's story of the two disciples in their shared meal with the risen Jesus at Emmaus (Lk. 24.25-35). The Eucharistic meal strengthens them, clarifies the perplexity that they shared earlier with the stranger who walked with them, and gives them the courage to return to Jerusalem and to their companions, who had also experienced the presence of the Risen Jesus in their declaration about the appearance of Jesus to Simon Peter (Lk. 24.34).

(g) A final note concludes the first part of this nautical event: after the meal by which Paul's travelling companions are encouraged and strengthened with a renewed sense of God's protection, 'they' cast the grain cargo into the sea in their effort to lighten the ship's weight (27.38b). Who the 'they' stands for is unspecified. The whole group aboard the ship or the sailors only? What 'they' do, though, is ecologically significant.

Taking our cue from earlier stories in the gospel, the contemporary listener might be reminded of Jesus' double exorcisms in Lk. 8.22-39. The first is the storm on the lake of Galilee that threatens to annihilate Jesus' disciples in the boat which is in danger of sinking (Lk. 8.22-25), a situation not too dissimilar to that which Paul is facing in 27.27-38.[16] The disciples awaken the sleeping Jesus, unperturbed by what is happening. He exorcizes the 'wind and the raging waves'. The storm ceases and a 'great calm' results (Lk. 8.24). The 'calm' indicates that Jesus has returned the water, the primordial Earth element, to its original paradisaical state.[17] This occurs through his exorcizing the evil present in the wind and the waters, other aspects of Earth's creation, that are not of themselves malevolent but have been invaded by evil. While there is no act of exorcism in the Acts' story, the Eucharistic meal provides an antidote to the fear experienced on Paul's ship caused through the evil responsible for the winds that batter the vessel. In Acts, the water is not the cause of the problem about which Paul and his companions are concerned; it is the wind. The whole sailing journey to this point has been governed by the direction and strength of the wind, whether favourable or not.

Luke's second exorcism story (Lk. 8.26-39) concerns one who is possessed by many demons. He becomes the quintessential human being devoid of community, care and protection. Jesus' response is to exorcize the evil spirits that have possessed the person and restore him to the fullness of his humanity.[18] Jesus does this through allowing the demons to enter a large herd of swine. The pigs rush down the slope of a steep bank

---

[16] On the theology of sea storms in Luke-Acts, see Charles Talbert, *Reading Luke-Acts in its Mediterranean Milieu* (Leiden: Brill, n.d.), pp. 175–95. Talbert sees Lk. 8.22-25 as an 'anticipation' or 'foreshadowing' of Paul's story in Acts (pp. 186–8, 192).

[17] See *AEC*, pp. 153–4.

[18] See *AEC*, pp. 155–6.

into the lake and are drowned in the lake's water, itself exorcized in the immediately preceding scene. The pigs, which have ecological value in the gospel, become the means for the man's rescue.[19] In a sense, through their drowning they become the sacrificial means that brings release.

Again, there is something similar in the action which closes the first part of Luke's dramatic story in Acts. The grain, the fruit of Earth, which has ecological value in the Greco-Roman world dependent on its growth and availability, is sacrificed. It, like the pigs of Luke's gospel, is poured into the waters to rescue a vulnerable vessel and its passengers. It is not clear how this 'release' will happen, at least not at this moment in the story. The ship is lightened, but what results in the second part of the story (Acts 27.39-44), seems disastrous.

'Saved on Earth' (27.39-44)

The second part of the Luke story (27.39-44) now begins with the evangelist's Earth-related observation, 'In the morning, they did not know the Earth (*gē*).' (Acts 27.39). At one level of listening, Luke could be suggesting that the sailing entourage 'did not recognize land' (NRSV), the usual translation. However, at an ecological level of listening, Paul's companions 'did not know the Earth (*gē*)'. They had become disconnected from Earth's possibilities of rescue and liberation. Their synchronicity with Earth had been compromised by the events that had befallen them and their preoccupation with self-rescue. They were now blind to Earth's presence and the possibility of rescue that it could offer. Not 'knowing' Earth reflects the ship's passengers' lack of intimacy with Earth through the vessel's inability to sail safely because of the power of the wind. 'Knowing' in the biblical sense implies intimacy.[20] Even in the presence of the morning light, they do not have this Earth-connectedness. This is how Luke begins this second part of the story: Those who sail are disconnected from Earth, which features two more times (27.43b, 44) before the story concludes. By the story's end, their relationship to Earth changes, but not without struggle and suffering and the possibility of their drowning.

They first intend to make for a bay with a beach (27.39). They cast the ship's anchors, slacken the ropes that secure the rudder and hoist the sail (27.40). But their efforts are thwarted. The vessel strikes a shoal, runs aground and breaks up (27.41). The ship, like all the other vessels that Paul uses in his various journeys, is an ecologically sophisticated means of transport created from wood, iron, rope and fabric. The creation of human ingenuity, it is Earth's gift of transportation to benefit the human community. It, too, like the wheat, becomes a sacrificial offering for the waters.

These two Earth images, of grain and a crushed sinking vessel, eventually allow the rescue of the ship's passengers. Some swim to shore; others, protected by the centurion

---

[19] About the gospel's pigs, see *AEC*, pp. 156.
[20] Dru Johnson, *Biblical Knowing: a Scriptural Epistemology of Error* (Eugene, OR: Wipf & Stock, 2013), pp. 23-7.

whose sailors wanted to kill, are ordered to throw themselves overboard and 'make for Earth (*gē*) and the rest on planks or on pieces of the ship. And so it happened that all were saved (*diasozō*) on (*epi*) the Earth (*gē*)' (27.43b-44).

Several features are ecologically significant here and confirmed by the particularity of Luke's language. The narrator notes those who throw themselves overboard and immerse themselves in water: the primordial Earth-symbol of creation, an event that might remind the contemporary listener of Jesus' water immersion in the early chapters of the gospel (Lk. 3.21-22). Jesus' water-immersion confirms his relationship to God as beloved and highly favoured son. Might this be a possibility in this water-immersion scene in Acts? This act releases them to 'make for Earth' as it confirms, through their connection to Earth, their relationship to God. Others come to Earth using 'planks and pieces of the ship', wood salvaged from the vessel. This other gift of Earth, wood debris, finally rescues those unable to swim. The result: 'all were saved (*diasozō*).'

The verb which Luke uses to indicate what happened to them all, that they were 'saved' (*diasozō*), has *sozō* as its Greek root. Already noted is how this theme of salvation expressed through *sozō* is already important in Luke's narrative (27.20, 31). The implication here, as the second part of Luke's story comes to its completion, is that Paul and his fellow voyagers are not simply rescued; they are 'saved', with all the theological import implied behind the use of this word already noted. The passive form of the verb ('were saved') indicates that this is God's act.

There is a further important point to note that is linked to the salvation which happens. They are all 'saved *on* (*epi*) the Earth' (27.44). Their restored connection to Earth (this is the sense of the preposition *epi* in Luke's text) becomes the confirming salvific experience. Disconnection from Earth, caused through the storm on the seas created by violent winds, has deepened their sense of insecurity. The fusion of Earth-related elements – the Eucharistic meal, the casting of grain into the sea, the fragile vessel which eventually runs aground, the waters themselves and the floating wood debris provided by the wrecked ship – all finally bring Paul's sailing companions to Earth. They are more than safe. They are *saved*. Their renewed ecological connectedness allows for the final part of Paul's journey to Italy and Rome to continue. This is the focus of Luke's final chapter in the Book of Acts, to which I now turn.

From Malta to Rome (28.1-31)

The specific expression of Earth to which the rescued arrive is the island of Malta (28.1). The local people receive Paul and his companions (including the 'us') with kindness, kindling a fire against the cold and rain (28.2). Paul's effort at collecting firewood results in him being bitten by a viper (28.3). In the literature of the classical world, the viper's attack would appear to be a sign of divine judgement.[21] Earth's creatures, in this case a viper, enact God's judgement on human beings. It would lead some to conclude that Paul is a murderer and expect his immediate demise (28.4). But

---

[21] Talbert, *Reading Luke-Acts*, pp. 182-3.

he shakes the viper off his hand. The creature does not bring God's condemnation but, instead, confirms a power that Paul has which will be put into action. This power overcomes evil and represents an 'authority to walk over snakes and scorpions, over every power of the enemy' (Lk. 10.18-19).[22] It is a power that Paul will exhibit to the end of Acts.

His unaffected condition convinces the local people he is, instead, a 'god' who becomes and acts, like Luke's Earth Child, as a healer (28.5-10). He cures the father of the bedridden hospitable island chief: 'Paul visited him and prayed, and putting his hands on him healed him.' (28.8b). This act of healing restores the man to wholeness in a manner that affirms that God's *basileia-ecotopia* continues to be active in Earth's children who faithfully replicate the ministry of Earth's Child. Paul's prayer prior to placing his hands on the father confirms that this is God's act, not Paul's. Paul is the agent of the *basileia-ecotopia*, not its inventor. What he does here in action, in revealing the presence of the *basileia-ecotopia*, will have its counterpart in the final verse of the Book of Acts, as Luke concludes his story of Earth's children noting Paul's preaching about God's *basileia-ecotopia* (28.31a), a point to which I shall return later. In other words, Paul continues to be a revealer of the *basileia-ecotopia* in deed and word until the end of Luke's wonderful story.

The success of Paul's healing action on the chief's father attracts the rest of the Maltese people who are also in need of healing (28.9). Their response is to honour their

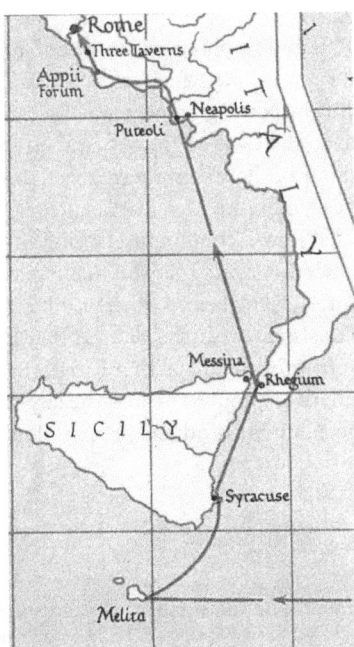

Illustration 15 Paul's final journey to Rome, via Malta and Puteoli.[23]

[22] Johnson, *Acts*, p. 462.
[23] Map from Orchard, *A Catholic Commentary*, p. 1312.

visitors with gifts and to resource adequately the vessel upon which Paul and the others leave from Malta to continue their journey to Rome after their three-month sojourn on the island (28.10-11). This delay would bring more favourable winds and allow them safe passage for their onward journey.[24]

From Malta, the group sails on a ship from Alexandria, Egypt (Illustration 15; 'with the Twin Brothers as its figurehead': 28.11c), to Syracuse, Rhegium, at the toe of Italy, finally arriving at the Italian port of Puteoli (28.11-13). The vessel's figurehead, 'Twin Brothers' (*Dioskuroi*), represented the twin sons of Zeus and Leda, Pollux and Castor, the navigators' protective deities who appeared as stars in the Mediterranean night sky.[25] The auditor would know that it is not these heavenly legendary deities who have control over the winds and seas and thus provide safe navigation. Rather, the Creator God, the deity in whom Paul believes, and by whose direction he journeys to Rome, is the true protector of the seas and all who sail upon them.

The mention by Luke of the *Dioskuroi* is not a superfluous detail but a reminder to auditors to reassess their commitment to the One that Paul preached about in Athens and who finally brings him to Puteoli, 'The God who made the *cosmos* and everything in it, being Lord of Heaven and Earth ... in whom we live and have our being' (17.24, 28).

Rome's Grain Port: Puteoli (28.11-13)

Puteoli was the major harbour and shipping transport hub for Rome, even eclipsing the importance of Rome's closer port, Ostia which, as noted in Chapter 10, welcomed two thousand ships a year, each transporting an average of almost 7,000 tonnes of grain.[26] Located near Herculaneum in the Gulf of Naples, in the region of Campania, Puteoli was a natural recessed harbour on Italy's southern coast, protected on its western side by a peninsula and on the east by Cape Misenum. A Roman colony from 194 BCE, it was a major port that serviced Rome, contemporary with the river port at Ostia.[27]

We know from many ancient writers that the main vessels that arrived at Puteoli carried Egyptian grain from Alexandria and other products from all over the Roman world.[28] This is certainly consistent with the picture that Luke paints of the type of vessel, a grain bearing ship from Alexandria, that carries Paul and his companions to the Italian mainland.[29] The harbour housed nearby the largest naval fleet in the ancient world, and was also an export hub for products produced in the Campanian region.[30]

---

[24] Dunn, *Acts*, p. 350.
[25] Dunn, *Acts*, p. 350.
[26] Temin, *Roman*, p. 40.
[27] Livy, *Ab Urbe Condita*, 34.45; Strabo, *Geographica*, 5.4.245; Vell. Pat., *Historiae Romanae*, 1.15.
[28] Cic., *Rab. Post.* 40; Suet., *Aug.* 98; Strabo, *Geographica*, 17.793; Cic., *Cael.* 10.
[29] James Smith, *The Voyage and Shipwreck of St. Paul with Dissertations on the Life of St. Luke, and the Ships and Navigations of the Ancients* (London: Longmans, Green and Co., 1880), pp. 156-7, offers a comprehensive picture of the sailing situation at the harbour of Puteoli.
[30] Fausto Zevi, 'Le grandi navi mercantili, Puteoli e Roma,' *Publications de l'École Française de Rome* 196 (1994), pp. 61-8.

Thus, Paul arrives at a final port in the story of Acts on the last leg of his journey to Rome at a maritime city that was prosperous, with a high level of private wealth, many luxurious villas, several granaries for the reception and storage of grain, and a harbour, given its imperial importance, constantly maintained by the emperors.[31] This is more than a stopping off port for Paul to visit local Jesus companions for seven days (28.14a) before moving on to Rome. For the contemporary auditor, it is an ecologically redolent urban maritime centre that brings together all the environmental implications that have accompanied Earth's children throughout Acts. Paul's 'seven day' stay at Puteoli would imply that he gathered with the local household of Jesus disciples for a celebration of the 'breaking of bread' on the 'first day' of the week.[32]

As the auditor knows from the gospel, this gathering would celebrate the fruits of Earth that become food for the community of Jesus followers.[33] The celebration of the 'breaking of bread', an event that Paul enacted not too many verses earlier to strengthen his disheartened travelling companions (27.33-36), would also be a source of strength for the Jesus household at Puteoli. It would also remind them about the importance of their ascetical style of discipleship that would release them of their possessions for the good of all. Taking the theme from the style of household discipleship in the early chapters of Acts, where Jesus members held all things in common, it would not be too hard to imagine that Paul's words to his companions ('brothers and sisters': 28.15a) at Puteoli would be to live simply, celebrating the gifts of creation and the God of Heaven and Earth, a theme that has accompanied Luke's Paul throughout Acts.

From Puteoli, Paul and his friends would have journeyed the rest of the way to Rome along the famous fourth-century BCE Appian Way. The image of Paul and friends walking this part of the journey by foot along an ancient route would cement for the auditor the physical and ecological implications of Paul's journey to Rome, the ancient and central city in the Roman Empire. This journey led him in a criss-crossing manner from Jerusalem, to Damascus, throughout Asia Minor, Macedonia, Greece, Israel, and across the seas of the Mediterranean. Luke concludes, 'And so to Rome we came.' (28.14c). En route to Rome, though, Luke mentions how members of the Roman Jesus households met them first at the Forum of Appius, and then at the 'Three Taverns' (28.15). Are these two different groups of Jesus followers? If so, what is their relationship to each other and why don't they meet them as one? Perhaps they reflect the diversity of Jesus households that Paul addressed in his Letter to the Romans, written some two decades earlier. Combining the information implicit in Romans with Luke's account, founded in historical memory, Paul's reaction to seeing both groups could well be one that 'thanked God and took courage' (28.15d).

---

[31] John D'Arms, 'Puteoli in the Second Century of the Roman Empire: a Social and Economic Study,' *The Journal of Roman Studies* 64 (1974), pp. 104–24, especially pp. 120–1.

[32] Paul's Sabbath day gathering with the Jesus followers at Puteoli is not beyond the realms of possibility. In fact, it would be most probable. Josephus mentions the presence of a Jewish community there, one of the oldest in the diaspora (*Ant.*, 17.328).

[33] See *AEC*, pp. 196, 199, 217; Lk. 14.7-28.

In Rome, Paul is allowed to live privately but under house arrest (28.16). Three days after he arrives he calls together the Roman Jewish leadership and, in another *apologia* contrary to the judgement of some, reaffirms his commitment to his Jewish heritage. One reason for coming to Rome was to meet the Jewish leadership, reaffirm his Jewish heritage and to reassure them of his commitment. Their response assures Paul that they have heard nothing that would contradict Paul's position or from anyone who has spoken evil against him. However, they recognize that the 'sect' that Paul advocates meets criticism 'everywhere' (28.17-22). This comment provides Luke's Paul with a platform to sum up the Christocentric teaching that has typified his preaching throughout Acts. With Paul's teaching Luke concludes this epic story of Earth's children.

Paul's Ecological Dwelling (28.23-28)

A great number come to Paul's dwelling, itself an ecologically integral space in which he can summarize the story of Earth's Child continued through Earth's children (28.23a). This is the story of God's action revealed first through the Torah and the Prophets (23.23b). Paul's preaching divides his audience: some accept it, others refuse (28.24). His final words in Acts draw on Isa. 6.9-10. They repeat Luke's adaptation of Isaiah in the gospel after Jesus' meta-parabolic teaching on the seed (Lk. 8.10 which draws on Isa. 6.9) and add an extra verse from the prophet (Isa. 6.10), shaping it in a way that interprets the situation with which Paul and, presumably, Luke's household wrestle.

> [26]'Go to this people and say,
> You will indeed listen, and never understand,
> and seeing you will see, but never perceive.
> [27]For the heart of this people has become hardened,
> and their ears hear with difficulty,
> and they have shut their eyes;
> lest they see with their eyes,
> and listen with their ears,
> and understand with their heart and turn,
> and I would heal them.'
>
> 28.26-27

In the gospel, the prophet offers Luke a way of understanding the reason that Jesus' teaching seems enigmatic to some. In Acts, the adaptation of Isaiah provides the evangelist with a way of understanding why most Jews did not follow Jesus and become members of the Lukan Jesus household. Luke's conclusion is clear: it was divinely foretold through Isaiah, that God's salvation would to come to the Gentiles: 'Let it be known to you then that this salvation of God has been sent to the Gentiles. They will listen.' (28.28).

It must be noted that in Luke's story of Earth's children, despite the apparent rejection of Paul's message throughout Acts by his Torah abiding co-religionists, the Jewish people are not excluded, even though at this stage in the narrative – at the end of Acts – the turn to the Gentiles is again reaffirmed. More is yet to come. The history of Jewish-Christian relations, while not overwhelmingly integral and affirming, has, in recent years, been seriously revised by contemporary Jesus followers. Those who now hear Luke's stories of the Jews in the gospel and Acts rethink the evangelist's perspective. They affirm and reclaim a new relationship, as sons and daughters of Abraham and Sarah.[34]

Witnessing to the End of the Earth (28.30-31)

Paul's final words and deeds in a household context continue to reveal the presence of the *basileia-ecotopia* in Rome. It might seem that he has finally arrived at the 'end of the Earth' located in the urban centre of the Roman Empire. However, as indicated in Chapter 2, 'end of the Earth' can also be a multivalent ecological metaphor for the contemporary auditor living with environmental sensitivity and asceticism to ensure that Earth does not end but is preserved and cared for. This spirit can also be gleaned in the final verse of Acts. Luke's concluding note is one of ecological optimism. This is evident in Luke's description of Paul's ascetical freedom which can be heard freshly by contemporary auditors, Earth's present children: 'He lived there [in Rome] for a whole two years at his own expense and welcomed all who came to him, proclaiming the *basileia-ecotopia* and teaching about the Lord Jesus Christ with all boldness (*parrēsia*) and without hindrance (*akōlutōs*).' (28.30-31).

Paul lives with a spirit of material freedom and hospitality, the essential virtue for those who care for the Earth. He lives 'at his own expense' for two years (28.30). This could mean several things: that he lived by himself, in a temporary dwelling which he rented himself.[35] How he paid for this rent for two years, as Luke envisages it, is conjectural. Had he accumulated enough money from his *skēnopoios*-tent-making to cover his costs? Did he have supporters and patrons?[36] His style of living in Rome was not avaricious, rather, he demonstrates a freedom from greed and a receptivity to all who visit him. Hospitality to all Earth's creatures and beings becomes the means by which the *basileia-ecotopia* is proclaimed and the teachings of Jesus honoured. Paul's style of living, and this teaching focus, continue with 'all boldness and without hindrance' (28.31).

The auditor might hear these two notes that conclude Acts, 'all boldness (*parrēsia*) and without hindrance (*akōlutōs*)' in several ways. In Luke's world *parrēsia* was the virtue of civic courage in expressing truth.[37] It has a social function that involves the

---

[34] This new reappreciation of the relationship between Judaism and Christianity was initially inspired in the Catholic Church by the Second Vatican Council, *Nostra Aetate: the Declaration of the Relation of the Church with Non-Christian Religions* (Vatican: Vatican City, 1965).
[35] Stanton, 'Accommodation', p. 243.
[36] Stanton, 'Accommodation', p. 244.
[37] Heinrich Schlier, 'παρρησία', *TDNT* 5, pp. 871–86, esp. pp. 871–2. See also, Gerhard Delling, 'Das letzte Wort der Apostelgeschichte', *NovT* 15 (1973), pp. 193–204.

wider community in which the person exercising *parrēsia* lives. The act of *parrēsia* also has moral consequences within the civic context. It brings others into the sphere of truth expressed by the bearer of *parrēsia*. Earlier in Acts, Luke uses this expression to characterize the witness of Peter and John before the Jerusalem religious authorities. Afterwards the Jesus householders gather with Peter and John after their release and pray for the gift of *parrēsia* in their ongoing preaching (4.29). God responds to their request, as noted in Chapter 4: 'And when they prayed the place in which they were gathered shook, and all, being filled with the Holy Spirit, spoke the Word of God with boldness (*parrēsia*).' (4.31).

Within the context of Paul's final teaching, as the ecologically oriented auditor hears it, the *parrēsia* that Paul expresses involves his conviction about the *basileia-ecotopia* and Jesus: Earth's Child. These two themes – the ecologically rich tensive symbol expressed in the *basileia-ecotopia* explored over Luke's two volumes, and Luke's Christology of Earth's Child which continues to be present in Acts – are the final focus of Paul's Roman preaching. These two themes are ecologically significant. They reveal God's attitude and presence on the Earth with Earth's creatures, expressed in the gospel through Jesus' ministry and, in Acts, through Earth's children.

The second note that describes Paul's preaching and the final word which concludes Acts is 'without hindrance' (*akōlutōs*). This implies that Paul's preaching about the *basileia-ecotopia* and Earth's Child continues, receives a welcome audience and, presumably, influences his listeners. If this preaching has ecological resonances, it focuses on the creator God whose presence and action confirm the ongoing presence of the *basileia-ecotopia*: a reality and encounter which Luke's audience would know is already present and 'in their midst'. For the ecologically attuned auditor who remembers Luke's gospel, it would give confidence for the future, that Earth will continue to be the subject of God's action and the agent that cooperates in God's work. Paul's unhindered preaching would further mean that 'the end of the Earth' will not necessarily come about because of the environmental destruction that this image might project. Earth can continue to be the focus of God's loving focus and action. However, Paul's preaching will also mean that this can only be assured because of its protection and care from Earth's children. This will require *parrēsia*.

Conclusion

The final two chapters of Acts have Paul undertake his journey, his final one, to Rome. These chapters are critical. They bring Paul to the geographical end-point towards which everything in Acts has been working. To the auditor, Paul's effort to get to Rome almost feels like his arrival at what might appear as 'Earth's end'. Luke's description of the voyage is dramatic, filled with accurate navigational details. The kinds of vessels that Paul and his companions travel on and the locations of harbours to which they arrive, reveal an intimate knowledge of maritime movement around the Mediterranean. This might reflect the author's accurate historical memory of Paul's travels or an awareness of the classic genres of voyages and shipwrecks, well described by ancient writers, that the writer incorporates into Paul's final voyage.

Whatever the background the author presumes, the auditor is gripped by the narrator's description of what occurs and is easily drawn into the story-world of Paul's final journey, especially everything that surrounds the shipwreck and how he completes the last part of his voyage to Italy. Already detected, though, are rich ecological themes that underpin the story. For the contemporary auditor, Paul appears in these closing chapters of Acts as an advocate of Earth's care by what he says and does:

- He encourages his travelling companions, distraught by their storm-tossed vessel, to 'remain in the boat'. His words reflect the importance of communal support and the network of Earth's children acting together to bring about liberty and safe passage, not only for themselves, but also for Earth's 'end'.
- He invites those who accompany him on the voyage, twice acknowledged as dangerous (27.9,10) into shared table communion through 'food' (*trophē*) that has salvific implications for its participants (27.33-38).
- The Eucharistic food in which all share provides the strength for them to release those things that weigh the ship, its tackle and eventually the grain cargo. These are sacrificial acts by Earth's gifts, reminiscent of gospel stories where similar events occur, to allow the rescue of the human cargo. Finally, the ship itself is sacrificed.
- Only then, when the wreck of the ship finally occurs, do its passengers reconnect to Earth. Symbolically they have been distanced from it. Assisted by the ship's timbers, the final group of passengers and crew, make it safely 'on' Earth (27.44). This reconnection assures their salvation, a prominent theme in this section of Acts.
- Paul becomes the advocate of the *basileia-ecotopia*. In the gospel his is the divine revealing action of God within all of creation, human and non-human. It is a tensive fruitful symbol of God's intention for creation revealed on Earth, first through the gospel ministry of Earth's Child and continued in Acts through Earth's children, the members of the Jesus movement. In these final chapters it is explicitly taught and enacted through Paul. He heals the sick and infirm, and teaches and advocates for its ongoing presence in a final scene in Rome, in a domestic setting: an ecologically symbolic space.
- As Acts concludes, Paul preaches this advocacy for the *basileia-ecotopia* with 'all boldness (*parrēsia*) and without hindrance (*akōlutōs*)'. Luke concludes on an optimistic note, highlighting Paul's courage and fidelity to the mission entrusted to him as he brings God's message, the story of Jesus, to 'the end of the Earth'. For ecological auditors this means that his audience, in Luke's own day and ours, are entrusted to speak the message in whatever place or 'end of the Earth' they find themselves.
- Finally, there is another point to consider about Earth's 'end'. Luke brings Paul to Rome. But this is not the end of the story. Earth's 'end' might be the city of Rome to which Paul finally arrives, preaching 'with boldness and without hindrance' (28.31). However, with an environmental sensitivity that the contemporary auditor brings, Earth's end would also refer to everything associated with Earth that is marginal and fragile. It is a realization that the Earth needs care. The injunction from the risen and ascended Jesus to witness to Earth's end (1.8c) continues to be an invitation to live with environmental awareness and ecological asceticism that would characterize Earth's children.

Part Four

CONCLUSION

Luke's Ecological Resonances in Acts

I began this commentary on Acts with an environmental snapshot of what has happened in Australia. I conclude with a report from the United Nations that judges a storm that has ravaged southern Africa, as the 'worst weather-related disaster ever to hit the southern hemisphere, with 1.7 million people in the path of the cyclone in Mozambique and 920,000 affected in Malawi ... storm surge floods up to 6 meters deep had caused "incredible devastation" over a huge area ... The Buzi river had burst its banks, killing hundreds'.[1] There is a growing urgency within the global community to respond to these ecological crises that continue to become more critical. The summary of the seventh session of the United Nations sponsored Intergovernmental Science-Policy Platform on Biodiversity and Ecosystem Services (IPBES), further reinforces this urgency. The summary makes for sobering reading:

> (A)round 1 million animal and plant species are now threatened with extinction, many within decades, more than ever before in human history. The average abundance of native species in most major land-based habitats has fallen by at least 20%, mostly since 1900. More than 40% of amphibian species, almost 33% of reef-forming corals and more than a third of all marine mammals are threatened. The picture is less clear for insect species, but available evidence supports a tentative estimate of 10% being threatened. At least 680 vertebrate species had been driven to extinction since the 16th century and more than 9% of all domesticated breeds of mammals used for food and agriculture had become extinct by 2016, with at least 1,000 more breeds still threatened.[2]

This response to this urgent ecological and environmental crisis, which rigorous authoritative scientific evaluation affirms, must also come from those of us involved in theological and biblical education. This present work is one response that seeks to support the efforts of Christian theologians and biblical scholars to undergird the

---

[1] Tom Miles, 'U.N. says 1.7 million in path of cyclone in Mozambique,' *Reuters, Environment*, 19 March 2019, https://in.reuters.com/article/us-africa-cyclone-un/u-n-says-1-7-million-in-path-of-cyclone-in-mozambique-idINKCN1R013J (accessed 21 March 2019).
[2] The summary of the seventh session of the United Nations sponsored Intergovernmental Science-Policy Platform on Biodiversity and Ecosystem Services (IPBES) further reinforces this urgency: see https://www.ipbes.net/ (accessed 21 March 2019).

invitation to ecological conversion. In this spirit I have sought to bring an ecological consciousness to a 'listening' to the Acts of the Apostles.

As mentioned throughout, its author was not an environmentalist as we understand this today. But through an intertextual approach – allowing our personal 'text' shaped by the present ecological concerns that surround us to encounter the 'text' of Luke's ancient work – we can allow ecological resonances to emerge from the story. While these were not intended by the evangelist, we cannot escape their implications for us now. As listeners to Luke's story and concerned about the natural world in which we live, we cannot remain neutral or passive interpreters.

The Acts of the Apostles is the second of two works written by the person who authored the Gospel According to Luke. The issues that faced the audience addressed by both writings remained the same: social stratification that caused division amongst later Jesus followers. Luke appeals to the elite of the Jesus household culturally and chronologically distant from the originating Galilean story of Jesus of Nazareth. The appeal is to release themselves from economic power and control to create a household of inclusivity for all: women and men, slaves and free, elite and poor. Luke's portrait of Jesus, as the prophet whose agenda is to bring about liberation and freedom for all (Lk. 4.16-21), and his teaching on discipleship free from material avariciousness (Lk. 12.15-34) urges a conversion of heart amongst the primary, though not exclusively elite audience addressed by Luke-Acts. This agenda continues through Acts. Ecological and environmental implications can be observed in each section of Acts.

Acts in Summary

1. An Ecological Orientation (1.1-5)

Luke links to the gospel with a prologue (1.1-5) similar to the one that began the gospel. These opening verses also reintroduce ecological themes familiar from the gospel: the Word, the presence of God's 'reign' (the *basileia-ecotopia*), and the Holy Spirit. Of central importance is Luke's confirmation of the communion between Heaven and Earth, a key theme in the gospel, and a motif which will be present throughout Acts in different reiterations.

2. The Ecological Mission (1.6-11)

The Earthly yet Risen Jesus gives his disciples a final directive that shapes the way the auditor follows the rest of Acts. Jesus confirms that they will receive 'power' from God – a rich ecological theme – which will enable them to be Jesus' witnesses 'in Jerusalem and in the whole of Judea and Samaria and to the end of the Earth (*gē*)' (1.8). This directive explicitly identifies Earth's presence for the first time and the movement which Jesus' witnesses will undertake over *gē*'s surface. What the meaning of 'end of the Earth' looks like will become clearer as the auditor moves towards the end of Luke's narrative in Acts' final chapters. Jesus ascends into Heaven. Earth's Child continues to be present to Earth's children and communion between Heaven and Earth is assured.

3. The Ecologically Renewed Household (1.12–2.47)

The image of the Jerusalem Jesus gathering in the 'upper room' echoes the Risen Jesus' communion with his disciples on Easter day and his encouragement for them to be his witnesses. The Twelve are reconstituted as (in the ecologically redolent day of Pentecost) the Spirit descends upon them. Peter's speech which follows (2.14-36) interprets the Spirit's action through the prophet Joel when 'all flesh' (2.16) will prophesy. This includes Earth as one of the participants that 'speaks' God's prophetic voice. Repentance follows, and Luke concludes with a summary of the qualities that characterize this renewed Jesus household (2.42), with its members committed to sharing Earth's resources (2.44).

4. The Fruitfulness of Earth's Children (3.1–6.7)

The leaders of the renewed Jerusalem Jesus household, Peter and John, minister in a way that reflects the healing deeds of Jesus in the gospel. They heal one unable to walk (3.9) as Peter speaks to his Jerusalem audience about the divine source of the healing act. After their release from those who want to punish them, their prayer of thanksgiving acknowledges the God of Creation, 'who made the Heaven and the Earth and the seas and everything in it' (4.24). This affirmation of God as Creator will find its echo later in the prayer-theology of Luke's Paul. Luke again affirms the unity of the Jesus household, its witness to the Risen Jesus (4.32-37), and their freedom in sharing Earth's resources. This spirit contrasts to greed that still resides amongst Jesus followers (5.1-11) and ethical tensions in Eucharistic practice as the Seven are appointed to address these issues (6.1-7). The Jerusalem Jesus household continues to flourish (6.7).

5. Earth's Presence in Stephen's Story of Israel (6.8–8.1a)

Two of the Seven, Stephen and Philip, begin to preach around and beyond Jerusalem. Stephen is martyred but not before Luke portrays him as the 'wonder worker' (6.8) and sums up Israel's history in explicit Earth (*gē*) terms (7.3, 4, 6, 29, 33, 36, 40, 49). Israel's story is an ecological one that confirms again the communion between Heaven and Earth. At the death of Stephen those who stone him, 'laid their garments at the feet of a young man named Saul' (7.58). This introduces us to Saul who will occupy most of Luke's final part of Acts. It also introduces us to the role which clothing, Earth's garments, plays, explicitly and implicitly, in Acts.

6. Water and Earth (8.1b–9.31)

Stephen's death signals a new wave of persecution against Jesus followers. Some flee throughout Judea and into Samaria (8.1b-3). The apostles remain in Jerusalem. Philip preaches to the Samaritans, and assists an Ethiopian returning from Jerusalem to understand the meaning of the Scriptures he is reading, finally baptizing him (8.4-40). The ecological motifs that surround this event anticipate the expansion of the Jesus movement beyond Jerusalem and the Jewish world. Saul experiences his encounter with the Risen Jesus on his trek to Damascus to capture and imprison Jesus followers to return them to Jerusalem (9.1-30). His experience is Earth-connected and is complete in his own baptism in Damascus by Ananias. Saul's journey to Damascus allows us to pause to reflect on the geographical and topographical impact on Jesus followers, like Saul, who would have walked everywhere as they travelled over the lands surrounding the Mediterranean.

7. Earth's Linen Sheet (9.32–11.18)

Peter travels outside Jerusalem towards the coast, a theological image of the impending endorsement by the leader of the embrace by members of the Jesus movement of the Gentiles. He heals two people in situations evocative of Jesus' gospel healings and outside the expected Jewish environment (9.35-43). This prepares Peter for a roof-top vision of a heavenly Earth-sheet as he lodges with the ritually unclean Simon, a tanner (10.9-16). The perplexing vision becomes clarified when Peter meets the Roman centurion, Cornelius, and his household later in the coastal city of Caesarea (10.24-29). Cornelius' witness to God's presence through prayer and obedience (10.30-32) brings Peter to recognize the action of God amongst the Gentiles and he baptizes them (10.44-47). His action is later confirmed by the Jerusalem Jewish leaders (11.1-18) and a new moment in the story of Acts unfolds. All this is instigated by Peter's vision of Earth's linen sheet and Earth's creatures that the sheet bears.

8. Earth's Interconnectivity and the God of Creation (11.19–14.28)

The stage is set for Saul's return into Luke's storyline. Saul's reintroduction is prefaced with Agabus' prophecy about the 'great famine (*limos*) over the whole world (*oukoumenē*)' (11.28). A worldwide ecological and economic crisis offers the context for Saul's future mission. This situation finds an echo later when Luke notes of the royal treatment of the country (*chora*) and the lack of availability of food to foreigners (12.20d). In the meantime, back in Jerusalem, James is executed by Herod who has Peter placed in prison (12.1-11). His release offers a further commentary on the role which Earth's iron and clothing play in Luke's story. Saul (now called Paul) and Barnabas begin their first missionary journey, sailing to Cyprus and on to Pamphylia in southern Asia Minor. This first of many of Paul's sailing excursions provides an opportunity to reflect on the ecological and environmental consequences from ships, their construction, passengers and the (usually elite) motivation that urged intense cargo passage around the Mediterranean. The preaching of Paul and Barnabas succeeds in allowing the Lord's word to 'spread abroad throughout the whole of the region (*chora*)' (13.49). It affects the environment in which they move. Paul's speech to the Lystrans praises the God of creation who is ecologically abundant (14.15-17).

9. Earth acts at Philippi (15.1–16.40)

Jerusalem Jesus members come to Antioch and demand that Paul and Barnabas justify their conduct among the Gentiles (15.1). They go to Jerusalem where their mission to the Gentiles is endorsed (15.2-21) and communicated back in Antioch (15.22-33). A second missionary journey takes Paul overland through Asia Minor and across to Macedonia. His Sabbath meeting of Lydia and her female companions at Philippi's river (16.13), a highly evocative environmental context, brings about Paul's first European baptism. Contrasting events that respect Earth, in the imagery that surrounds Lydia and her occupation, or abuse Earth, especially in Paul's treatment at the hands of the Philippian authorities (16.22), lead to the conversion of the jailer (16.27-34). He offers hospitality to Paul and his preaching companion, Silas, who finally leave Philippi after bidding farewell to Lydia (16.38-40). Earth is present and acts throughout the whole Philippian event.

10. The God of Life and Breath (17.1–18.1)

Paul and Silas continue their journey through Europe. A mixed reaction follows Paul's preaching in the Thessalonian Synagogue (17.2-7). The two are accused of turning 'the world (*oikoumēnē*) upside down' (17.6-7). This accusation is ecologically true, as Jesus' disciples seek to live with an awareness of justice and shared resources in a Roman world controlled by imperial politics and economics that supports the wealthy and elite. In Athens, Paul offers one of the most important ecologically explicit speeches in Acts as he appeals to Stoics and Epicureans in their respective philosophical traditions about life and the physical world (17.6-23). Paul reiterates his faith in a God of the cosmos who is 'Lord of Heaven and Earth ... gives to all things, life and breath and everything ... God exists not far from each one of us. For in God we live and move and have our being' (17.24-8). Paul's speech is a highpoint in Luke's affirmation of the God of Creation.

11. The Artisan, Artemis and the Lord's Supper (18.2–20.12)

Paul comes to Corinth. The auditor learns of Paul's artistry as a skilled (*technē*) tent-maker (18.2c-3). This skill puts him into contact with Earth and animal skins or goat hair. Sailing to Ephesus, Paul finds himself in conflict with the silversmiths of the Artemis cult who financially benefit from the cult (19.23-41). In Troas, Paul gathers with Jesus disciples on the 'first day of the week' for the 'breaking of the bread' (20.7). Paul's Eucharistic practice sustains him and his companions. It provides the setting for Paul's resuscitation of the young Eutychus who went into a deep sleep from Paul's long speech and fell three storeys to his death (20.9-12).

12. Earth's Child identifies with Earth's Children (20.13–26.32)

Paul meets with the Ephesian elders at Miletus (20.17-38), encouraging them in their shepherding of Jesus followers. In his speech, Luke's Paul reveals the ascetical generous spirit that guides his mission and conduct (20.33-35). Paul then sails to Caesarea for a short stay (21.7-9), goes to Jerusalem and addresses the criticism that his antagonists have levelled against him (22.1-21). Paul's address contains ecological themes already familiar to the auditor. It highlights his encounter with the Risen Jesus and the role which light plays in the experience, and reiterates the communion between Heaven and Earth. Finally, the Roman tribune rescues Paul from a violent mob and has him furtively transferred to Caesarea (23.12-35).

13. The Final Voyage towards Rome and Earth's 'End' (27.1–28.31)

The final two chapters of Acts tell the story of Paul's most daring and dangerous voyages that bring him eventually to Italy and to the story's denouement, Rome, the heart of the Roman Empire and a controlling ecological urban epicentre. Behind Luke's story is the auditor's understanding of shipping and sea travel that the author incorporates into the dramatic narrative, discussed earlier in Chapter 8. Ecological themes from the gospel and earlier chapters of Acts find their summary. These include salvation, Earth-connectedness and dependency, food and the household space. Acts concludes with Paul proclaiming about *basileia-ecotopia* and Jesus' teaching 'with all boldness (*parrēsia*) and without hindrance (*akōlutōs*)' (28.30-31). Paul has arrived, in one sense, at Earth's ends: Rome. The auditor knows, however, that Earth's ends may never come. Its communion with Heaven is assured and repeated throughout Acts. Further, Luke's final two notes in Paul's address to his Roman Jewish audience, *parrēssia* and *akōlutōs*, bring hope for the environmental future of an Earth looked after by Jesus' disciples imbued with a generous spirit of an ascetical lifestyle shaped by environmental conversion and practice.

Concluding Ecological Insights

Environmental resonances emerge from the perspective of the intertextual ecological hermeneutic that can influence the way the contemporary auditor hears Luke's story in Acts. Luke's teaching from Earth's Child finds its expression in the principal characters that dot the chapters of Acts. These are Earth's children. Ecological themes noted in the gospel continue into Acts.

In Acts, Luke reaffirms that ongoing presence of God's reign, of the *basileia-ecotopia*. This is the cultural realization of God's intended communion upon the Earth amongst all Earth's creatures. This communion between Earth and Heaven is again acknowledged in the ascension of Earth's Child (1.9-10) whom the 'two men' remind his gazing disciples will return 'as you have beheld him go into Heaven' (1.11). The communion and interpenetrability of the two spheres of cosmic existence, Earth and Heaven celebrated in the gospel, revealed through the birth and ministry of Earth's Child and in Jesus' ascension at the beginning of Luke's second volume, remain throughout Acts. In Acts, Earth becomes the arena of God's encounter with humanity. Creation, Earth's tangible expression, is the place of this encounter and God's self-revelation. God is Lord of Heaven and Earth. This is confirmed in the prayer offered on the release of

Peter and John after their detention by the Jerusalem authorities (4.23-30). They address the God of Heaven and Earth. Paul's later speech to the Athenians also acknowledges God as 'Lord of Heaven and Earth' and the source of all life and our very being (17.24-28). God acts and is present in creation.

The principal agent that affirms this Earth-Heaven communion and assists people to know God's life is the Holy Spirit, first promised by the Risen Jesus to his gathered disciples on Easter night (Lk. 24.48-49). At the ecologically rich festival of Pentecost this Spirit descends, unites and empowers Jesus disciples to witness to God. The Spirit authorizes them to realize Jesus' injunction in 1.8, to be his witnesses 'in Jerusalem, and in the whole of Judea and Samaria and to the end of the Earth (*gē*)'. This witness occurs in many ways as Luke's story of the movement of Earth's children in the rest of Acts unfolds.

Earth's presence accompanies them throughout:

- The Jerusalem Jesus gathering occurs within a domestic setting, the 'upper room', an ecologically defined space with strong Earth-connected memories from the gospel, as they await the coming of Jesus' promised Spirit (1.13). Their style of communal living in which no one was in want among them, and their freedom to release themselves of property and possessions (2.42-47; 4.32-37) reveal Luke's idyllic resentation of the Jesus household living out the teaching of Jesus. A similar space concludes Luke's story of Paul as he speaks to his Roman audience about God's *basileia-ecotopia* and confirms the original teaching of Jesus (28.30-31). This domestic ecological space frames the whole of Acts.
- Earth (*gē*) features in Stephen's summary of Israel's salvation in a long speech to his Jerusalem executioners about to stone him to death (7.2-53). Its presence appears throughout Acts, presumed or in its elements.
- Water is an actor in the baptism of the first non-Jerusalemite (8.4-40). This act by Earth's primordial substance through the agency of Philip on an Ethiopian, prepares for the eventual mission to the wider Greco-Roman world.
- A linen sheet, a product of Earth, comes down from Heaven before Peter in a perplexing vision that finally gets its resolution in his baptism of a Gentile family (10.9-47).
- Earth's iron and clothing further participate in Luke's story. They are present in times of incarceration and release (12.1-11; 16.22-34).
- Earth's ground accompanies Paul throughout all his overland missionary journeys, as he walks everywhere, is subject to the elements and the contours of the land and lives a humble existence seeking to support himself as an artisan skilled in tent-making (18.2c-3).
- Earth's seas are also integral to the journeys of Paul and his companions. While he walks everywhere, for reasons of convenience to move quickly from one part of the Mediterranean to another, he also travels by ship. As noted, the ships that Paul travels in are themselves the product of ancient human ingenuity constructed out of Earth's products available to their builders. There is nothing artificial about these vessels. They are subject to Earth's elements of water and wind. What happens to

them on the seas is dependent on winds, whether favourable or storm-filled. They also carry human beings, though their owners, for economic reasons tinged with attitudes of greed, prefer cargo and grain, as they help reinforce the Roman imperial economic system.

The above represent only some of the Earth-affirming insights that have emerged from this ecologically attentive listening to the Acts of the Apostles. These intuitions may not have been the original intent of its author, but this ancient text can continue to influence and shape the world in which we, listeners and disciples of Jesus, live. These insights or intuitions might be transferred into six theses that could deepen the spirit of ecological conversion amongst contemporary Jesus followers.

1. *The story in Acts is an Earth-story.* As the story of Jesus' gospel is ecologically connected, the story of the disciples in Acts is about their encounter with Earth, Earth's people, cities and lands. The intertextual shift from an anthropocentric interpretation of Acts to an environmentally oriented 'listening' allows Earth's presence to surface and become an actor in the mission entrusted to the disciples by the Risen Jesus in 1.8.
2. *Heaven and Earth are in communion with each other.* The separation between Earth and Heaven is overcome in the ministry of Luke's Jesus. This communion continues into Acts. This means that God's presence occurs on Earth and within the experiences of Jesus' witnesses, in which Earth's elements and gifts feature. This communion also means that Earth and Jesus followers are cared for and watched over by Earth's risen and ascended Child whose presence empowers the disciples through the Holy Spirit.
3. *God is Lord of Heaven and Earth.* At significant moments in Acts, and especially in prayer, the disciples acknowledge God as Lord of Heaven and Earth (4.23-30; 17.24-8). This further confirms the communion between Earth and Heaven, that these are not separate spheres of existence. The same God is equally present to both and holds both in care.
4. *Earth acts.* Acts unfolds in an environment that is not just a literary world of Luke's imagination. The Earth is not simply a backdrop for Luke's story of the Jesus movement. The real world and natural environment participate in the story. Whether Luke's characters walk, sail, baptize, clothe themselves or become imprisoned, Earth accompanies them, participates in their witness and collaborates with them.
5. *Earth is a revealer of the sacred.* This is clear in Paul's speech to the Athenians (17.22-9). His presentation in this one scene communicates Luke's theology of the natural world. The Earth, the natural environment in which people live, becomes the first moment of their encounter with God, however inarticulately this encounter is perceived.
6. *Members of the Jesus movement are Earth's Children.* Discipleship in Acts is not environmentally neutral. Those who witness to Jesus are ascetically committed. They do not accumulate Earth's wealth or goods. They generously offer what they have to others; they model the virtues of hospitality and food sharing because they are daughters and sons of Earth.

These six theses summarize the ecologically oriented insights that emerge from a particular mode of attentiveness to Luke's second volume. They cohere with the seven theses from *About Earth's Child* that affirm the ecological connectedness of Luke's Jesus and the gospel disciples who act with a spirit of ecological conversion and environmental asceticism.[3] This is the heart of Jesus' teaching. This spirit continues into Acts. The commissioned disciples witness to Jesus' teaching and the *basileia-ecotopia* in their mission towards 'the end of the Earth'. Luke's two-volumed work, the gospel and the Acts of the Apostles, can powerfully deepen the spirit of ecological conversion for contemporary disciples, today's Earth children, concerned about the environment and the natural world that surrounds them.

---

[3] *AEC*, pp. 302-4.

Bibliography

Armstrong, Karl, 'A New Plea for an Early Date of Acts,' *Journal of Greco-Roman Christian Judaism* 13 (2017), pp. 79–110.
Ascough, Richard S., *Lydia: Paul's Cosmopolitan Hostess* (Collegeville, MN: Liturgical Press, 2009).
Babie, Paul, and Michael Trainor, *Neo-Liberalism and the Biblical Voice: Owning and Consuming* (New York and London: Routledge, 2018).
Bain, Katherine, *Women's Socioeconomic Status and Religious Leadership in Asia Minor in the First Two Centuries CE* (Minneapolis: Fortress Press, 2014).
Balz, Horst, 'οἰκουμένην,' *EDNT* 2, eds Horst Balz and Gerhard Schneider (Grand Rapids, MI: Wm B. Eerdmans Publishing Company, 1991), pp. 503–4.
Balz, Horst, and Gerhard Schneider (eds), *EDNT* (3 vols.; Grand Rapids, MI: William B. Eerdmans Publishing Company, 1990).
Barrett, C. K., *A Critical and Exegetical Commentary on the Acts of the Apostles: Volume 2, Introduction and Commentary on Acts XV-XXVII* (London: T. & T. Clark, 1998).
Barrett, C. K., *Luke the Historian in Recent Study* (London: Epworth Press, 1961).
Beutler, J., 'μάρτυς,' *EDNT* 2, eds Horst Balz and Gerhard Schneider (Grand Rapids, MI: Wm B. Eerdmans Publishing Company, 1991).
Bird, Michael F., 'The Unity of Luke-Acts in Recent Discussion,' *JSNT* 29 (2007), pp. 425–48.
Blue, Bradley, 'Acts and the House Church,' *The Book of Acts in Its First Century Setting, Volume 2: Graeco-Roman Setting*, eds David W. J. Gill and Conrad Gempf (Grand Rapids, MI: William B. Eerdmans Publishing Company, 1994), pp. 119–222.
Bornkamm, Gunther, 'σεισμός; σείω,' *TDNT* 7, eds Gerhard Kittel and Gerhard Friedrich (Grand Rapids, MI: Wm B. Eerdmans Publishing Co., 1964), pp. 196–200.
Brinks, C. L., '"Great is Artemis of the Ephesians": Acts 19.23-41 in Light of Goddess Worship in Ephesus,' *CBQ* 71 (2009), pp. 776–94.
Britner, Bradley, 'Acclaiming Artemis in Ephesus: Political Theologies in Acts 19,' *The First Urban Christians 3: Ephesus*, eds James R. Harrison and L. L. Welborn (Atlanta: SBL Press, 2018), pp. 127–70.
Britt, John, 'All about Iron,' *Ceramics Monthly* 59 (2011), pp. 14–15.
Brown, Colin (ed.), *NIDNTT* 1 (Exeter: Paternoster Press, 1975).
Bruce, F. F., *The Book of Acts* (Grand Rapids, MI: Wm. B. Eerdmans Publishing Company, 1988).
Campbell, Joan Cecilia, *Phoebe: Patron and Emissary* (Collegeville, MN: The Liturgical Press, 2009).
Campbell, William S., *The 'We' Passages in the Acts of the Apostles: the Narrator as Narrative Character* (Atlanta: Society of Biblical Literature, 2007).
Carter, Warren, and Amy-Jill Levine, *The New Testament: Methods and Meanings* (Nashville: Abingdon Press, 2013).
Clay, Diskin, 'The Athenian Garden,' *The Cambridge Companion to Epicureanism*, ed. James Warren (Cambridge: Cambridge University Press, 2009), pp. 9–28.

Conzelmann, Hans, *Acts of the Apostles: A Commentary on the Acts of the Apostles* (Minneapolis: Fortress Press, 1987).
Crowther, Nigel, 'Visiting the Olympic Games in Ancient Greece: Travel and Conditions for Athletes and Spectators,' *The International Journal of the History of Sport* 18 (2001), pp. 37–52.
D'Arms, John, 'Puteoli in the Second Century of the Roman Empire: A Social and Economic Study,' *The Journal of Roman Studies* 64 (1974), pp. 104–24.
Delling, Gerhard, 'Das letzte Wort der Apostelgeschichte,' *NovT* 15 (1973), pp. 193–204.
Delling, Gerhard, 'πλεροω,' *TDNT* 6, eds Gerhard Kittel and Gerhard Friedrich (Grand Rapids, MI: Wm B. Eerdmans Publishing Co., 1964), pp. 286–311.
Derks, Hans, '"The Ancient Economy": The Problem and the Fraud,' *The European Legacy* 7 (2002), pp. 597–620.
Dockery, David S., 'Acts 6–12: The Christian Mission Beyond Jerusalem,' *Review and Expositor* 87 (1990), pp. 423–37.
Dschulnigg, Peter, 'Die Rede des Stephanus im Rahmen des Berichtes über sein Martyrium (Apg 6,8-8,3),' *Judaica* 44 (1988), pp. 195–213.
Duling, Dennis, *The New Testament: an Introduction* (New York: Harcourt, Brace, Jovanovich, 1993).
Dunn, James D. G., *The Acts of the Apostles* (London: Epworth Press, 1996).
Dupont, Jacques, 'Le douxième apôtre (Acts 1.15-26): à propos d'une explication récente,' *The New Testament Age: Volume 1*, ed. W. C. Weinrich (Macon, GA: Mercer University Press, 1984), pp. 139–45.
Dupont, Jacques, *The Salvation of the Gentiles* (New York: Paulist Press, 1967).
Edwards, Denis, *Breath of Life: A Theology of the Creator Spirit* (Maryknoll, NY: Orbis Books, 2004).
Edwards, Denis, *Christian Understandings of Creation: The Historical Trajectory* (Minneapolis: Fortress, 2017).
Edwards, Denis, *Deep Incarnation: God's Redemptive Suffering with Creatures* (Maryknoll, NY: Orbis Books, 2019).
Edwards, Denis, *Ecology at the Heart of Faith* (Maryknoll, NY: Orbis Books, 2008).
Edwards, Denis, *How God Acts: Creation, Redemption, and Special Divine Action* (Minneapolis: Fortress, 2010).
Edwards, Denis, *Jesus and the Cosmos* (Eugene, OR: Wipf & Stock, 2004).
Edwards, Denis, *Jesus and the Natural World: Exploring a Christian Approach to Ecology* (John Garrett: Melbourne, 2012).
Edwards, Denis, *Made from Stardust: Exploring the Place of Human Beings within Creation* (North Blackburn, Vic.: Collins Dove, 1992).
Edwards, Denis, *The Natural World and God: Theological Explorations* (Hindmarsh: ATF Press, 2017).
Eliot, T. S., *Selected Essays, 1917–1932* (NY: Harcourt, Brace & Company, 1932).
Ellis, E. Earle, '"The End of the Earth" (Acts 1:8),' *BBR* 1 (1991), pp. 121–32.
Elvey, Anne, *An Ecological Feminist Reading of the Gospel of Luke: A Gestational Paradigm* (Lewiston, NY: Edwin Mellen Press, 2005).
Fischer-Hansen, Tobias, and Birte Poulson, *From Artemis to Diana: The Goddess of Man and Beast* (Acta Hyperborea 12. Copenhagen: Museum Tusculanum University of Copenhagen Press, 2009).
Fitzgerald, Michael, 'The Ship of Saint Paul, Pt 2: Comparative Archaeology,' *BA* 3 (1990), pp. 31–9.

Fitzmyer, Joseph, *Acts of the Apostles: A New Introduction with Introduction and Commentary* (Yale: Yale University Press, 1998).
Fitzmyer, Joseph, 'The Ascension of Christ and Pentecost,' *TS* 45 (1984), pp. 409–40.
Freyer-Griggs, Daniel, 'The Beasts at Ephesus and the Cult of Artemis,' *HTR* 106 (2013), pp. 459–77.
Friedrich, G., 'Dunamis,' *EDNT* 1, eds Horst Balz and Gerhard Schneider (Grand Rapids, MI: William B. Eerdmans Publishing Company, 1990), pp. 355–8.
Friesen, Steven, 'Poverty in Pauline Studies: Beyond the So-Called New Consensus,' *JSNT* 26 (2004), pp. 323–61.
Gilbert, Gary, 'The Acts of the Apostles.' *The Jewish Annotated New Testament: New Revised Standard Version*, eds Amy-Jill Levine and Marc Zvi Brettler (Oxford: Oxford University Press, 2011), pp. 197–9.
Gill, David W. J., and Conrad Gempf (eds), *The Book of Acts in Its First Century Setting, Volume 2: Graeco-Roman Setting* (Grand Rapids, MI: William B. Eerdmans Publishing Company, 1994).
Gill, David, 'Acts and the Urban Elites,' *The Book of Acts in Its First Century Setting, Volume 2: Graeco-Roman Setting*, eds David W. J. Gill and Conrad Gempf (Grand Rapids, MI: William B. Eerdmans Publishing Company, 1994), pp. 105–18.
Goppelt, Leonhard, 'πεινάσω (λιμὸς),' *TDNT* 6, eds Gerhard Kittel and Gerhard Friedrich (Grand Rapids, MI: Wm B. Eerdmans Publishing Co., 1964), pp. 12–22.
Green, Joel B., 'The Death of Jesus and the Rending of the Temple Veil (Luke 23:44-49): A Window into Luke's Understanding of Jesus and the Temple,' *The Society of Biblical Literature 1991 Seminar Papers*, ed. Eugene H. Lovering Jr (Atlanta, GA: Scholars Press, 1991), pp. 543–57.
Green, Joel B., *The Theology of Luke* (Cambridge: Cambridge University Press, 1995).
Habel, Norman C. (ed.), *Readings from the Perspective of Earth* (Sheffield: Sheffield Academic Press Ltd, 2000).
Habel, Norman C., 'Introducing Ecological Hermeneutics,' *Lutheran Theological Journal* 46.2 (2012), pp. 97–105.
Haenchen, Ernst, *The Acts of the Apostles: A Commentary* (Philadelphia: The Westminster Press, 1965).
Hampson, Margaret R., *A Non-Intellectualist Account of Epicurean Emotions* (MPhil, London: University College London, 2013).
Harrison, James R., and L. L. Welborn (eds), *The First Urban Christians 3: Ephesus*, (Atlanta: SBL Press, 2018).
Hirschfeld, Nicole, 'The Ship of Saint Paul, Pt 1: Historical Background,' *BA* 53 (1990), pp. 25–30.
Hock, Ronald F., *The Social Context of Paul's Ministry: Tentmaking and Apostleship* (Philadelphia: Fortress Press, 1980).
Hooker, Morna, 'Artemis of Ephesus,' *JTS* 64 (2013), pp. 37–46.
Horrell, David G., 'The Label Χριστανοζ: 1 Peter 4.16 and the Formation of Christian Identity,' *JBL* 126 (2007), pp. 361–81.
Houston, Walter, 'What Was the Meaning of Classifying Animals as Clean and Unclean,' *Animals on the Agenda: Questions about Animals for Theology and Ethics*, eds Andrew Linzey and Dorothy Yamamoto (London: SCM Press Ltd, 1998), pp. 18–24.
Hurtado, Larry W., 'Convert, Apostate, or Apostle to the Nations: The "Conversion" of Paul in Recent Scholarship,' *Studies in Religion* 22 (1993), pp. 273–84.
Ikas, Karin, and Gerhard Wagner (eds), *Communicating in the Third Space* (Routledge Research in Cultural and Media Studies 18. London: Taylor & Francis, 2009).

Jensen, Lloyd B., 'Royal Purple of Tyre', *Journal of Near Eastern Studies* 22, no. 2 (1963), pp. 104–18.
Johnson, Dru, *Biblical Knowing: A Scriptural Epistemology of Error* (Eugene, OR: Wipf & Stock, 2013).
Johnson, Luke Timothy, *The Acts of the Apostles* (Collegeville, MN: The Liturgical Press, 1992).
Jones, Brice C., 'The Meaning of the Phrase "And the Witnesses Laid Down Their Cloaks" in Acts 7.58,' *Expository Times* 123 (2011), pp. 113–8.
Judge, Edwin A., 'The Social Identity of the First Christians: A Question of Method in Religious History,' *Journal of Religious History* (1980), pp. 201–17.
Karris, Robert, *Works of St. Bonaventure: Commentary on the Gospel of Luke. Chapters 1–8* (Saint Bonaventure, NY: Franciscan Institute Publications, 2001).
Keener, Craig S., *Acts: An Exegetical Commentary, Volume 2* (Grand Rapids, MI: Baker Academic, 2013).
Kidder, Annemarie S. (ed.), *The Mystical Way in Everyday Life: Sermons, Essays and Prayers*: Karl Rahner, S. J. (Maryknell, NY: Orbis Books, 2010).
Kilgallen, John J., 'The Function of Stephen's Speech (Acts 7.2-53),' *Biblica* 70 (1989), pp. 173–93.
Kim, Seyoon, *Paul and the New Perspective: Second Thoughts on the Origin of Paul's Gospel* (Grand Rapids, MI: William B. Eerdmans Publishing Co., 2002).
Kittel, Gerhard, and Gerhard Friedrich (eds), *Theological Dictionary of the New Testament*, translated by Geoffrey W. Bromiley (10 vols.; Grand Rapids, MI: Wm B. Eerdmans Publishing Co., 1964–76).
Kochenash, Michael, 'Better Call Paul "Saul": Literary Models and a Lukan Innovation,' *JBL* 138 (2019), pp. 433–9.
Légasse, Simon, *Stephanos: histoire et discours d'Étienne dans les Actes des Apôtres* (Lectio Divina 147. Paris: Cerf, 1992).
Levenson, Jon, *Creation and the Persistence of Evil* (Princeton: Princeton University Press, 1988).
Levine, Amy-Jill, and Marc Zvi Brettler (eds), *The Jewish Annotated New Testament: New Revised Standard Version* (Oxford: Oxford University Press, 2011).
Linzey, Andrew, and Dorothy Yamamoto (eds), *Animals on the Agenda: Questions about Animals for Theology and Ethics*, (London: SCM Press Ltd, 1998).
Lohfink, Gerhard, *Jesus of Nazareth: What He Wanted, Who He Was* (Collegeville, MN: The Liturgical Press, 2012).
Lohr, Joel N., 'He Identified with the Lowly and Became a Slave to All: Paul's Tent-Making as a Strategy for Mission,' *Currents in Theology and Mission* 34 (2007), pp. 179–87.
Lolos, Yannis, 'Via Egnatia after Egnatius: Imperial Policy and Inter-regional Contacts,' *Mediterranean Historical Review* 22 (2007), pp. 273–93.
Lovering, Eugene H. Jr (ed.), The Society of Biblical Literature 1991 Seminar Papers (Atlanta, GA: Scholars Press, 1991).
Lüdemann, Gerd, *The Acts of the Apostles: What Really Happened in the Earliest Days of the Church* (New York: Prometheus Books, 2005).
Marshall, I. Howard, 'Brothers Embracing Sisters?,' *The Bible Translator* 55 (2004), pp. 303–10.
Marshall, I. Howard, *Luke: Historian and Theologian* (Exeter: Paternoster, 1970).
Marshall, I. Howard, *The Acts of the Apostles* (Sheffield: Sheffield Academic Press, 1992).
Menzies, Robert, 'Spirit and Power in Luke-Acts: a Response to Max Turner,' *JSNT* 15 (1993), pp. 11–20.

Metzger, Bruce, *A Textual Commentary on the Greek New Testament* (London: United Bible Societies, 1975).
Michel, Otto, 'ἡ οἰκουμένην,' *TDNT* 5, eds Gerhard Kittel and Gerhard Friedrich (Grand Rapids, MI: Wm B. Eerdmans Publishing Co., 1964), pp. 157–9.
Miller, Amanda C., 'Cut from the Same Cloth: A Study of Female Patrons in Luke–Acts and the Roman Empire,' *Review & Expositor* 114 (2017), pp. 203–10.
Minear, Paul, 'Dear Theo: the Kerygmatic Intention and Claim of the Book of Acts,' *Union Seminary Review* 27 (1973), pp. 131–50.
Murphy-O'Connor, Jerome, 'Travelling Conditions in the First Century: On the Road and on the Sea with St. Paul,' *Bible Review* 1 (1985), pp. 38–42, 44–7.
Murphy-O'Connor, Jerome, *Paul: a Critical Life* (Oxford: Oxford University Press, 1996).
Nanos, Mark D., and Magnus Zetterholm (eds), *Paul within Judaism: Restoring the First Century Context to the Apostle* (Minneapolis: Fortress Press, 2015).
Njoroge wa Ngugi, J., 'Stephen's Speech as Catechetical Discourse,' *Living Light* 33 (1997), pp. 64–71.
Orchard, Bernard, Edmund F. Sutcliffe, Reginald C. Fuller, and Ralph Russell (eds), *A Catholic Commentary of Holy Scripture* (London: Thomas Nelson and Sons Ltd, 1953).
O'Toole, Robert F., 'Philip and the Ethiopian Eunuch (Acts Viii 25–40),' *JSNT* 17 (1993), pp. 25–34.
Parsons, Mikeal C., *Acts* (Grand Rapids, MI: Baker Academic, 2008).
Parsons, Mikeal C., and Martin M. Culy, *Acts: a Handbook on the Greek Text* (Waco, TX: Baylor University Press, 2003).
Pervo, Richard I., *Acts: A Commentary* (Minneapolis: Fortress Press, 2009).
Pervo, Richard I., *Dating Acts: Between the Evangelists and the Apologists* (Santa Rosa, CA: Polebridge, 2006).
Petersen, David, *The Acts of the Apostles* (Grand Rapids, MI: Wm B. Eerdmans Publishing Co, 2009).
Pilch, John, *Visions and Healing in the Acts of the Apostles: How the Early Believers Experienced God* (Collegeville, MN: Liturgical Press, 2004).
Porter, Stanley, 'Excursus: the "We" Passages,' *The Book of Acts in Its First Century Setting, Volume 2: Graeco-Roman Setting*, eds David W. J. Gill and Conrad Gempf (Grand Rapids, Michigan: William B. Eerdmans Publishing Company, 1994), pp. 545–74.
Rahner, Karl, 'A Faith that Loves the Earth,' *The Mystical Way in Everyday Life: Sermons, Essays and Prayers: Karl Rahner, S.J.*, ed. Annemarie S. Kidder (Maryknoll, NY: Orbis Books, 2010), pp. 52–8.
Rapske, Brian, 'Chapter 1: Acts, Travel and Shipwreck,' *The Book of Acts in Its First Century Setting, Volume 2: Graeco-Roman Setting*, eds David W. J. Gill and Conrad Gempf (Grand Rapids, MI: William B. Eerdmans Publishing Company, 1994), pp.1–48.
Ravens, David, *Luke and the Restoration of Israel* (JSNT Supplement Series 119. Sheffield, UK: Sheffield Academic Press, 1995).
Rhoades, David, 'Performance Criticism: an Emerging Methodology in Biblical Studies,' https://www.sbl-site.org/assets/pdfs/Rhoads_Performance.pdf (accessed 12 February 2019).
Rohrbaugh, Richard L., 'The Social Location of the Marcan Audience,' *BTB* 23 (1993), pp. 114–27.
Rohrbaugh, Richard L., 'The Social Location of the Markan Audience,' *The Social World of the New Testament: Insights and Models*, eds Jerome H. Neyrey and Eric C. Stewart (Peabody Mass: Hendrickson Publishers, 2008), pp. 143–62.

Rollinson, Hugh R., *Early Earth Systems: A Geochemical Approach* (Malden, MA: Blackwell Pub., 2007).

Rossing, Barbara R., 'Trajan's Column and the Cargo List of Rev 18:12-13: John's Critique of Rome's Economy in Ecological and Eschatological Perspective,' presentation at the *Society of Biblical Literature* meeting, San Antonio, 2016.

Rutkin, Aviva, 'Iron Rain Left Heavy Metal on Early Earth,' *New Scientist* 225 (2015), p. 14.

Salles, Ricardo (ed.), *God and Cosmos in Stoicism* (Oxford: Oxford University Press, 2009).

Sand, Alexander, 'σάρξ,' *EDNT* 3, eds Horst Balz and Gerhard Schneider (Grand Rapids, MI: Wm B. Eerdmans Publishing Co., 1991), pp. 230–3.

Sasse, Hermann, 'γῆ, ἐπίγειος,' *TDNT* 1, eds Gerhard Kittel and Gerhard Friedrich (Grand Rapids, MI: Wm B. Eerdmans Publishing Co., 1964), pp. 677–81.

Sasse, Hermann, 'κοσμέω, κόσμος, κόσμιος, κοσμικός,' *TDNT* 3, eds Gerhard Kittel and Gerhard Friedrich (Grand Rapids, MI: Wm B. Eerdmans Publishing Co., 1964), pp. 867–98.

Scheidel, Walter, and Sitta von Reden, *The Ancient Economy* (New York and London: Routledge, 2002).

Schlier, Heinrich, 'παρρησία,' *TDNT* 5, eds Gerhard Kittel and Gerhard Friedrich (Grand Rapids, MI: Wm B. Eerdmans Publishing Co., 1964), pp. 871–86.

Schmidt, Thomas E., 'The Penetration of Barriers and the Revelation of Christ in the Gospels,' *NovT* 34 (1992), pp. 229–46.

Schwartz, Daniel R., 'The End of the Γῆ (Acts 1:8): Beginning or End of the Christian Vision?,' *JBL* 105 (1986), pp. 669–76.

Schweizer, Eduard, 'σάρξ,' *TDNT* 7, eds Gerhard Kittel and Gerhard Friedrich (Grand Rapids, MI: Wm B. Eerdmans Publishing Co., 1964), pp. 98–151.

Scott, James, 'Luke's Geographical Horizon,' *The Book of Acts in Its First Century Setting, Volume 2: Graeco-Roman Setting*, eds David W. J. Gill and Conrad Gempf (Grand Rapids, MI: William B. Eerdmans Publishing Company, 1994), pp. 483–545.

Second Vatican Council, *Nostra Aetate: the Declaration of the Relation of the Church with Non-Christian Religions* (Vatican: Vatican City, 1965).

Sellars, John, *Stoicism* (New York and London: Routledge, 2006).

Shiell, William, *Reading Acts: The Lector and The Early Christian Audience* (Leiden: Brill, 2004).

Simpson, D. P., *Cassell's New Latin-English English-Latin Dictionary* (London: Cassell, 1962).

Sleeman, Matthew, *Geography and the Ascension Narrative in Acts* (Cambridge: Cambridge University Press, 2009).

Smith, James, *The Voyage and Shipwreck of St. Paul with Dissertations on the Life of St. Luke, and the Ships and Navigations of the Ancients* (London: Longmans, Green and Co., 1880).

Smith, William, *Concise Dictionary of the Bible* (London: John Murray, 1900).

Soards, Marion L., *The Speeches in Acts: their Content, Context, and Concerns* (Louisville, Kentucky: Westminster/John Knox, 1994).

Soja, Edward W., *Postmodern Geographies: the Reassertion of Space in Critical Social Theory* (London: Verso, 1989).

Soja, Edward W., *Thirdspace: Journeys to Los Angeles and Other Real-and-Imagined Places* (Hoboken, NJ: Wiley, 1996).

Spicq, Ceslas, *Theological Lexicon of the New Testament* (Peabody, MA: Hendrickson, 1994).

Stanton, Greg, 'Accommodation for Paul's Entourage,' *NovT* 60 (2018), pp. 227–46.
Still, Todd D., 'Did Paul Loathe Manual Labour? Revisiting the Work of Ronald F. Hock on the Apostle's Tentmaking and Social Class,' *JBL* 125 (2006), pp. 781–95.
Strelan, Rick, 'Tabitha: The Gazelle of Joppa (Acts 9:36-41),' *BTB* 39 (2009), pp. 77–86.
Strømmen, Hannah M., *Biblical Animality after Jacques Derrida* (Atlanta: SBL Press, 2018).
Sylva, Dennis D., 'The Meaning and Function of Acts 7:46-50,' *JBL* 106 (1987), pp. 261–75.
Talbert, Charles, *Reading Acts: A Literary and Theological Commentary on the Acts of the Apostles* (New York: Crossroads, 1997).
Talbert, Charles, *Reading Luke-Acts in Its Mediterranean Milieu* (Leiden: Brill, n.d.).
Tannehill, Robert C., *The Narrative Unity of Luke-Acts: a Literary Interpretation, Volume 2: The Acts of the Apostles* (Minneapolis: Fortress Press, 1990).
Tannehill, Robert C., 'Israel in Luke-Acts: A Tragic Story,' *JBL* 104 (1985), pp. 69–85.
Taylor, Justin, 'Why Were the Disciples First Called "Christians" at Antioch,' *Revue Biblique* 101 (1994), pp. 75–94.
Temin, Peter, *The Roman Market Economy* (Princeton, NJ: Princeton University Press, 2012).
Trainor, Michael, *About Earth's Child: An Ecological Listening to the Gospel of Luke* (Sheffield: Sheffield Phoenix Press, 2012).
Trainor, Michael, *The Body of Jesus and Sexual Abuse: How the Gospel Passion Narratives Inform a Pastoral Response* (New York: Wipf & Stock, 2017).
Trudinger, Peter L., and Norman C. Habel, *Exploring Ecological Hermeneutics* (Atlanta: Society of Biblical Literature, 2008).
Turner, Max, *Power from on High: the Spirit in Israel's Restoration and Witness in Luke-Acts* (Sheffield: Sheffield Academic Press, 2000).
Van de Sandt, Huub, 'Why is Amos 5.25-27 quoted in Acts 7,42f.?,' *ZNW* 82 (1991), pp. 67–87.
Van der Horst, Pieter, 'Hellenistic Parallels to the Acts of the Apostles: 1.1-26,' *ZNW* 74 (1983), pp. 17–26.
Van Unnik, Willem C., 'Der Ausdruck ἙΩΣ ΕΣΧΑΤΟΥ ΤΗΣ ΓΗΣ (Apostelgeschichte I 8)....', *Sparsa Collecta, Volume 1*, eds J. Reiling, G. Mussies, P. W. van der Horst, L. W. Nijendijk (Leiden: Brill, 1973–83), pp. 386–91.
Vickers, Michael, 'Golden Greece: Relative Values, Minae, and Temple Inventories,' *American Journal of Archaeology* 94 (1990), pp. 613–25.
Warren, James (ed.), *The Cambridge Companion to Epicureanism* (Cambridge: Cambridge University Press, 2009).
Warren, James, *Epicurus and Democritean Ethics: an Archaeology of Ataraxia* (New York: Cambridge University Press, 2002).
Weinder, Francis D., 'Luke, Stephen, and the Temple in Luke-Acts,' *BTB* 17 (1987), pp. 88–90.
Weinrich, W. C. (ed.), *The New Testament Age: Volume 1* (Macon, GA: Mercer University Press, 1984).
Weins, Delbert L., *Stephen's Sermon and the Structure of Luke-Acts* (N. Richland Hills, TX: Bibal, 1996).
Weins, Delbert L., 'Luke on Pluralism: Flex with History,' *Direction* 23 (1994), pp. 44–53.
Wilson, S. G., 'The Ascension: A Critique and an Interpretation,' *ZNW* 59 (1968), pp. 269–81.
Winter, Bruce, 'Acts and Food Shortages,' *The Book of Acts in its First Century Setting, Volume 2: Graeco-Roman Setting*, eds David W. J. Gill and Conrad Gempf (Grand Rapids, MI: William B. Eerdmans Publishing Company, 1994), pp. 59–78.

Witherington, Ben, *The Acts of the Apostles: A Socio-Rhetorical Commentary* (Grand Rapids/Cambridge: Wm. B. Eerdmans, 1998).
Wycherley, Richard E., 'The Garden of Epicurus,' *Phoenix* 13 (1959), pp. 73–7.
Zerwick, Max, and Mary Grosvenor, *A Grammatical Analysis of the Greek New Testament, Volume 1: Gospels – Acts* (Rome: Biblical Institute Press, 1974).
Zevi, Fausto, 'Le grandi navi mercantili, Puteoli e Roma,' *Publications de l'École Française de Rome* 196 (1994), pp. 61–8.

Index of Authors

Armstrong, Karl 3, 173
Ascough, Richard S. 106, 173

Babie, Paul 40, 173
Bain, Katherine 117, 173
Balz, Horst 114, 173
Barrett, C. K. 3, 104, 173
Beutler, J. 30, 173
Bird, Michael F. 3, 173
Bornkamm, Gunther 108, 173
Bradley, Blue 9, 173
Brettler, Marc Zvi 44, 173
Brinks, C. L. 130, 173
Britner, Bradley 128, 173
Britt, John 87, 173
Brown, Colin 173
Bruce, F. F. 2, 173

Campbell, Joan Cecilia 22, 173
Campbell, William S. 104, 105, 173
Carter, Warren 44, 173
Clark, Stephen 33, 118
Clay, Diskin 117, 173
Conzelmann, Hans 2, 3, 38, 85, 88, 104, 175
Crowther, Nigel 69, 174
Culy, Martin M 3, 177

D'Arms, John 155, 174
Delling, Gerhard 134, 158, 174
Derks. Hans 4, 174
Dockery, David S. 52, 174
Dschulnigg, Peter 53, 174
Duling, Dennis 174
Dunn, James D. G. 79, 149, 154, 174
Dupont, Jacques 36, 148, 174

Edwards, Denis 13, 14, 15, 174
Eliot, T. S. 8, 174
Ellis, E. Earle 31, 174

Elvey, Anne 105, 174

Fischer-Hansen, Tobias 129, 174
Fitzgerald, Michael 97, 174
Fitzmyer, Joseph 33, 175
Flanagan, Richard 1
Freyer-Griggs, Daniel 129, 175
Friedrich, Gerhard 29, 173, 174, 175, 176, 177, 178
Friesen, Steven 118, 175
Fuller, Reginald C. 144, 177

Gempf, Conrad 4, 5, 9, 69, 85, 105, 177, 178, 179
Gilbert, Gary 44, 175
Gill, David W. J. 4, 5, 6, 9, 69, 85, 105, 175
Goppelt, Leonhard 85, 175
Green, Joel B 3, 28, 175 .
Grosvenor, Mary 85, 111, 179

Habel, Norman C. 2, 3, 60, 175
Haenchen, Ernst 2, 7, 175
Hampson, Margaret R. 118, 175
Harrison, James R. 128, 175
Hirschfeld, Nicole 97, 175
Hock, Ronald F. 124, 175
Hooker, Morna 129, 134, 175
Horrell, David G. 9, 175
Houston, Walter 82, 175
Hurtado, Larry W. 67, 175

Ikas, Karin 28, 175

Jensen, Lloyd B. 106, 107, 175
Johnson, Dru 151, 176
Johnson, Luke Timothy 2, 5, 27, 36, 38, 39, 53, 59, 60, 67, 72, 73, 125, 126, 130, 136, 143, 149, 153, 176
Jones, Brice C. 61, 176
Judge, Edwin A. 124, 176

Karris, Robert 79, 176
Keener, Craig S. 75, 76, 79, 80, 176
Kilgallen, John J. 52, 176
Kidder, Annemarie S. 14, 177
Kim, Seyoon 67, 176
Kittel, Gerhard 176, 177, 178
Kristeva, Julia 8
Kochenash, Michael 51, 176

Légasse, Simon 52, 176
Levenson, Jon 43, 176
Levine, Amy-Jill 44, 176
Linzey, Andrew 82, 176
Lohfink, Gerhard 36, 176
Lohr, Joel N. 123, 176
Lolos, Yannis 113, 176
Lovering, Eugene H. 28, 175
Lüdemann, Gerd 75, 78, 176

Marshall, I. Howard 3, 36, 148, 176
Menzies, Robert 29, 176
Metzger, Bruce 80, 90, 176
Michel, Otto 114, 176
Miller, Amanda C. 106, 176
Minear, Paul 45, 176
Murphy-O'Connor, Jerome 69, 70, 94, 95, 124, 145, 176, 177
Mussies, G 31

Nanos, Mark D. 67, 177

Orchard, Bernard 144, 153, 177
O'Toole, Robert F. 66, 177

Parsons, Mikeal C. 2, 3, 177
Pervo, Richard I. 2, 3, 4, 38, 105, 119, 177
Petersen, David 38, , 177
Pilch, John 75, 76, 79, 177
Porter, Stanley 105, 177
Poulson, Birte 129, 174

Rahner, Karl 14, 15
Rapske, Brian 69, 70, 94, 96, 97, 177
Ravens, David 53, 177
Reiling, J 31
Rhoads, David 9, 177
Rohrbaugh, Richard L. , 177
Rollinson, Hugh R. 87, 177

Rossing, Barbara R. 116, 177
Russell, Ralph 144, 177
Rutkin, Aviva 87, 177

Salles, Ricardo 119, 177
Sand, Alexander 38, 177
Sasse, Hermann 56, 60, 120, 177
Scheidel, Walter 4, 178
Schlier, Heinrich 158, 178
Schmidt, Thomas E. 59, 178
Schneider, G. 173
Schwartz, Daniel R. 10, 178
Schweizer, Eduard 38, 178
Scott, James 5, 178
Sellars, John 118, 178
Shiell, William 9, 178
Simpson, D.P. 7, 178
Sleeman, Matthew 28, 32, 33, 178
Smith, James 154, 178
Smith, William 93, 126, 178
Soards, Marion L. 38, 53, 178
Soja, Edward W. 28, 178
Spicq, Ceslas 119, 178
Stanton, Greg 75, 92, 125, 157, 178
Still, Todd D. 124, 178
Strelan, Rick 74, 75, 178
Strømmen, Hannah M. 82, 178
Sutcliffe, Edmund F. 144, 177
Sylva, Dennis D. 59, 178

Talbert, Charles 3, 150, 152, 178
Tannehill, Robert C. 2, 44, 47, 48, 53, 104, 125, 178, 179
Taylor, Justin 9, 179
Temin, Peter 114, 154, 179
Trainor, Michael 2, 40, 179
Trudinger, Peter L. 2, 179
Turner, Max 29, 179

Van de Sandt, Huub 58, 179
van der Horst, Pieter 19, 31, 179
van Unnik, Willem C. 31, 179
Vickers, Michael 96, 179
von Reden, Sitta 4, 178

Wa Ngugi, J. Njoroge 50, 177
Wagner, Gerhard 28, 175
Warren, James, 118, 175

Weinder, Francis D. 53, 179
Weinrich, W. C. 36, 179
Weins, Delbert L. 51, 53, 179
Welborn, L. L. 128, 175
Wilson, S. G. 33, 179
Winter, Bruce 85, 145, 179
Witherington, Ben 51, 53, 179

Wycherley, Richard E. 118, 179

Yamamoto, Dorothy 82, 175

Zerwick, Max 85, 108, 111, 179
Zetterholm, Magnus 67, 177
Zevi, Fausto 155, 179

Index of References

First Testament

Genesis		3.24	29	49.7-26	31
1	100, 105	16.9-12	37	53.7-8	67
				57.19	39
1.1	21	**Joshua**		58.6	89, 105
1.1-2.4	109	4.24	29	61.1	59, 89, 105
6.17	38				
11.31-32	54	**1 Kings**		66.1	54, 59
12.1	54	17.17-24	73		
12.1-4	54			**Jeremiah**	
15.13	58	**2 Kings**		16.21	29
16.1	54	4.32-7	73		
48	57			**Ezekiel**	
48.4	54, 57	**1 Chronicles**		33.4	125
		8.12	73		
Exodus				**Joel**	
2.22	54	**1 Maccabees**		2.17-21	
3	58	1.62-3	79	3.1-5	37
3.33	58	11.34	73		
12.11	88			**Amos**	
19.16-19	37	**Psalms**		2.25-27	58
20.11	100	63.2	38		
23	37	76.15	29	**Daniel**	
23.16	37	144.12	29	1.16-18	79
34.22	37			7.13	33
		Isaiah			
Leviticus		6.9	156	**Tobit**	
11	79	6.9-10	156	1.10-13	79
23.15-21	37	6.10	156		
		40.5	38	**Judith**	
Deuteronomy		49.1	31	9.8	29
2.5	54	49.6	31, 98	13.4	29

Second Testament

		1.3	19, 21, 22	1.26-38	6, 29
Mark		1.4	5	1.35	29, 37
1.9-12	24	1.5-23	22	1.39-45	23
		1.9	66	1.42	66
Luke		1.14	65	1.44	65
1-2	23, 24	1.22	66	1.69-71	148

1.77	148	8.22-39	150	16.19	91
2.1	114	8.24	150	16.19-31	6, 87, 91
2.7	9, 24, 35, 67, 74	8.26-39	5, 82, 150	16.21	91
		9.1-17	46, 66	17.20-27	22
2.9	88	9.10-16	147	19.9-10	148
2.10	5, 65, 67	9.10-17	49, 149	19.29	10
2.11	148	9.25	120	19.29-44	35
2.12	9, 24, 67	9.28	40	19.35	60
2.13	58	9.30	33	19.36	60, 61
2.14	24, 25, 32, 54, 57, 59, 62	9.34-5	33	19.38	25, 32, 59
		9.51–19.27	6, 9, 29	21.12	35
		10.3	134	21.18	148
2.16	9, 24, 67	10.8-9	22	21.37-38	35
2.30	148	10.17	65, 77, 78, 101	21.38	10
3.1-14	23			22.11	35
3.1-20	39	10.18-19	153	22.12	35
3.6	38, 148	10.20-22	40	22.14-23	49
3.15	23	10.21-22	40	22.15-30	10
3.16	23	10.38-42	48, 49	22.19	40
3.18-20	24	10.39-42	106	22.21-34	36
3.21	24, 40, 79	11.2	41	22.28-34	149
3.21-22	24, 29, 36, 66, 152	11.2-4	40, 136	22.39-46	35, 40
		11.3	41	22.39-53	10
4.1	40	11.4	41	22.47–23.25	136
4.1-30	130	11.50	120	23.6-12	10
4.6-11	130	12.13-21	46, 57, 96	23.26-49	10
4.13	66	12.15-34	164	23.50-53	10, 80
4.16-21	164	12.15-49	86	23.52	90
4.16-30	105, 106	12.16-21	6	23.53	74, 81
4.18	109	12.20	47	24.1	80, 81
4.18-19	88	12.20-21	47	24.1-12	67
4.18-21	37	12.22-34	148	24.3-4	80
4.25	86	12.30	120	24.4-5	33, 80
4.25-26	86	13.10-17	105	24.5	81
4.42-44	22	13.18	22	24.6	90
5.12-32	22	13.21-22	22	24.9	80
5.17-26	73	13.23–19.27	39	24.9-11	90
6.12	40	13.29	22	24.11	80
6.23	65	14.2-6	47	24.12	80, 81, 90
6.30	57	14.7-28	155	24.13-35	49
7.11-16	73	15	88	24.19	131
7.31-32	22	15.3-7	134	24.25-35	66, 150
8.3	57	15.7	65	24.28-29	149
8.4-21	50, 97	15.10	65	24.30	131
8.4-39	22	15.11-32	82, 86	24.30-31	49
8.10	156	15.14	86	24.34	81, 150
8.13	65	15.14-20	88	24.35	131
8.22-25	23, 145, 148, 149, 150	15.22	88	24.39	30
		16	22	24.41	65

24.41-42	30	2.14-36	37, 165	4.31	45, 158
24.42	49	2.16	165	4.32	47
24.45	30	2.17-18	37	4.32-35	47
24.47	26	2.17-21	37	4.32-37	42, 45, 50,
24.48-49	169	2.21	148		57, 91, 165, 169
24.49	37	2.29	45	4.34	47
24.50-53	19, 21, 22,	2.33	33	4.36-37	46
	36, 59	2.37-47	50	4.36–5.11	46
24.52	65	2.38	39	5.1-10	91
		2.38-47	39	5.1-11	20, 42, 46, 165
John		2.39	39, 73	5.12–6.7	48
1.14	105	2.40	148	5.4	46, 47
		2.41	39, 44	5.12-16	48, 50
Acts		2.42	39, 42, 72,	5.14	48
1.1	19, 25		131, 165	5.20	48
1.1-2	19	2.42-47	45, 169	5.17-39	48
1.1-5	12, 19, 25,	2.43	41	5.30-31	48
	26, 164	2.43-47	41	5.31	148
1.1–8.1	11, 12	2.44	41, 165	5.40	48
1.1–8.3	28	2.44-45	57	5.42	48, 50
1.3	19, 22	2.44-46	41	6	49
1.4-5	19, 22	2.46	41, 72, 131,	6.1	48
1.6	27		150	6.1-6	5, 42, 48
1.6-8	27	2.47	41, 148	6.1-7	51, 165
1.6-11	12, 27, 29,	3.1–6.7	12, 43, 45,	6.1–8.3	51
	31, 83, 164		47, 49, 165	6.2	48
1.7	27	3.1-26	61	6.4	49
1.8	10, 11, 28,	3.2-3	43	6.7	43, 49, 50, 52,
	29, 30, 31,	3.7-9	43		165
	69, 98, 160,	3.9	165	6.7–8.8	52
	164, 169,	3.12-26	38, 43, 98	6.8	51, 165
	170	3.13	43	6.8–8.1	12, 51, 55, 59,
1.9-10	168	3.15	43		61, 165
1.9-11	20, 32, 36,	3.17	43	6.8–7.58	61
	59	3.22	43	6.8–7.1	52
1.11	33, 168	3.26	44	6.8–8.1	52, 53
1.12-26	35	4.1-2	44	7	54, 58
1.12–2.47	12, 35, 165	4.1-23	44	7.2-19	54
1.13	35, 169	4.4	44	7.2-53	38, 52, 55, 169
1.14	36	4.8-12	45	7.3	55, 56, 165
1.15-22	36	4.12	148	7.3-6	54, 58
1.15-26	36	4.13	45	7.3-7	54, 55
2	34, 133,	4.21-22	45	7.5	57
	139	4.23-30	169, 170	7.4	55, 56, 165
2.1-13	20	4.24	45, 165	7.6	57, 62
2.1-36	36	4.24-35	45	7.7	57, 58, 59, 62
2.10	5, 6, 67	4.26	45	7.8	52
2.14-20	38	4.28	45	7.20-34	54
2.14-28	44	4.29	45, 158	7.29	56

7.33	54, 58, 62, 165	9.1-25	73	10.14	79
		9.1-29	136	10.14-15	79, 80
7.35-53	54	9.1-30	67, 166	10.15	79
7.36	56, 165	9.1-31	68	10.17	77, 78
7.40	56, 165	9.2	69, 136	10.17-48	101
7.42	58	9.3	71	10.19	77, 78
7.44	54	9.4	69, 70, 71	10.24-9	166
7.44-53	59	9.5-6	33, 70, 71	10.30-32	166
7.47	59	9.8	69, 70, 71	10.33	76
7.48	59	9.8-19	71	10.34	78
7.49	62, 165	9.15-16	72	10.34-43	38, 77
7.49-50	54, 58, 59, 60	9.17	71	10.44-46	77
		9.19	71	10.44-47	77, 101, 166
7.52	54	9.19-22	71	10.44-48	11, 109
7.54	44	9.23-25	71	10.47-48	78
7.54–8.1	52	9.26	71	10.48	23
7.55-56	33	9.27-29	71	11	78, 80, 83, 89, 90
7.56	60	9.29	71		
7.58	60, 64, 165	9.30	71	11.1-18	77, 101, 166
7.59-60	60	9.31	71	11.2-3	83
8	49, 66	9.32	39	11.5	140
8.1	11, 60, 165	9.32-35	11, 73	11.5-6	78, 82
8.1-3	65	9.32-43	101	11.5-10	77, 90
8.1-8	50, 52	9.32–11.18	12, 69, 73, 75, 77, 79, 81, 83, 166	11.12	78
8.1–9.31	12, 65, 71, 165			11.14	148
				11.15-16	83
8.1-26.32	12	9.34	33, 73	11.17	78, 83, 89
8.3	65	9.35-43	166	11.18	83, 85
8.4-40	65, 169	9.36-41	74	11.19-21	92
8.4–15.35	28	9.36-43	11	11.19-22	85
8.8	65	9.39	74	11.19-26	110
8.9-13	65	9.40	73, 74	11.19–14.28	12, 85, 87, 89, 91, 93, 95, 97, 99, 101, 166
8.14-17	65	9.41	74		
8.24	66	9.43	75		
8.26-39	5	10	68, 81, 82, 83	11.22-44	85
8.26-40	11, 23, 61, 73			11.25-46	85, 92
		10.1-2	75	11.26	9
	81, 101, 109	10.1-48	75	11.28	85, 101, 166
		10.1-11.18	49, 110	11.29	87
8.33	67	10.6	75	12	108
8.35	66	10.9-16	166	12.1-5	61
8.36	66	10.9-23	80	12.1-11	87, 109, 166, 169
8.38	67	10.9-47	169		
8.40	67	10.10	76	12.1-19	101
9.1	61	10.10-16	11	12.2	87
9.2	69, 137	10.11	80	12.4	87
9.1-2	68, 71	10.11-12	78	12.6	87
9.1-9	53	10.11-16	76	12.6-11	108, 110
9.1-19	70, 71	10.12	81		

12.7-11	88	15.1-29	75	16.27-29	109	
12.8	88, 89, 90	15.1-31	5	16.27-34	167	
12.10	88	15.1-35	49	16.30-31	148	
12.11	90	15.1–16.40	12, 102,	16.30-32	109	
12.18-19	90		105, 109,	16.33-34	109	
12.20	90, 91, 101, 166		110, 111, 167	16.34	109	
12.20-23	91	15.2-3	103	16.34-39	106	
12.23	91	15.2-21	167	16.35-36	110	
12.24	91	15.4	103	16.37	110	
13	100	15.5	103	16.38	106	
13.1-2	92	15.6-11	104	16.38-39	110	
13.1-49	91	15.6–16.10	104	16.38-40	167	
13.1–14.28	110	15.12	104	16.40	92, 106, 110	
13.2-3	92	15.13-21	104	16.44	109	
13.4	92	15.13-29	123	17.1	113	
13.4-6	92	15.20	136	17.1–18.1	12, 113, 115, 117, 119, 121	
13.4-13	23	15.21	106	17.2	44	
13.4–14.26	97	15.22-33	104, 167	17.2-3	113	
13.5-12	92	15.36	104	17.2-7	167	
13.9	92	15.36-40	104	17.4	113	
13.13	92	15.36–28.31	28	17.5	113	
13.13-47	97	15.37-38	92	17.6	114	
13.13-50	92	15.39	92	17.6-7	113, 167	
13.14	106	15.40-41	93	17.6-15	113	
13.16-41	38	16	109	17.6-23	167	
13.23	148	16.1-5	124	17.7	92	
13.26	148	16.1-10	104	17.9	113	
13.27	106	16.6–17.15	123	17.10-12	117	
13.42	44	16.11	23, 114	17.13-15	117	
13.45	98	16.11-12	93, 104	17.16-21	119	
13.46	98, 100	16.11-15	92	17.16-23	117	
13.47	31, 98, 101, 148	16.11-40	104	17.16-33	123	
		16.12-13	106	17.18	119	
13.48	98	16.12-40	106	17.19-20	119	
13.49	99, 166	16.13	105, 167	17.22-23	119	
14.1-6	99	16.14	106	17.22-29	170	
14.1-7	92	16.15	23, 106, 107	17.22-31	38	
14.1-28	99	16.16-18	106, 107	17.23	119	
14.15	101, 106	16.17	148	17.24	154	
14.15-17 166	99, 101,	16.19	107	17.24-28	120, 167, 169, 170	
		16.19-24	61			
14.19-23	100	16.22	61, 107, 167	17.26	122	
14.24-25	92	16.22-34	169	17.28	130, 154	
14.26	23, 92	16.23-4	108	17.29-31	121	
14.27	100	16.24-33	106	17.32	121	
15	20, 99, 111	16.25	108	17.34–18.1	121	
15.1	103, 167	16.25-34	89, 108	18-20	131	
15.1-12	123	16.26	108	18.1-2	123	

18.1-17	123	20.11	131	22.8	136		
18.2-3	123, 124, 131, 169	20.13	69	22.9	137		
		20.13-15	23	22.10	69, 137		
18.2–20.12	12, 123, 125, 127, 129, 131, 167	20.13-16	135	22.11	69		
		20.13–26.32	12, 133, 135, 137, 139, 167	22.11-13	137		
				22.14-15	33		
				22.14-16	137		
18.3	125	20.16	93	22.17-21	33, 137		
18.4	106, 125	20.17-35	133	22.22	137		
18.6	98, 125	20.17-38	133, 167	22.22-23	137		
18.6-7	125	20.18-21	133	22.23	137		
18.7-8	125	20.18-35	38	22.24	137		
18.9-10	125	20.19	139	22.25-29	137		
18.12-13	116	20.22-24	133	22.30	137		
18.12-17	126	20.24	134	23.1-9	137		
18.17	126	20.28	134	23.10	137		
18.18-22	23	20.29	129	23.11	138		
18.18-23	126	20.32	134	23.12-15	138		
18.21-3	93	20.33-35	134, 135, 168	23.12-35	168		
19.1	126			23.16-24	138		
19.1-7	127	20.35	139	23.23	156		
19.1–20.1	126, 133	20.36-38	135	23.25–25.8	138		
19.8-10	128	20.36–21.17	135	24	80		
19.9	69, 136	21.1-4	23	24.1-21	38		
19.10	132	21.3	135	24.5	114		
19.11-19	128	21.1-8	93	25.9-11	138		
19.20	128, 132	21.1-18	104	25.11-12	143		
19.21-2	128	21.5-6	135	25.12	138		
19.22–20.1	128	21.7-8	135	25.13	138		
19.23	69, 136	21.7-9	168	25.13–26.32	138		
19.23-27	128	21.8	23, 135	25.13–26.33	138		
19.23-41	130, 167	21.9	135	25.14-18	139		
19.26-27	132	21.10-12	136	26.2-23	38		
19.27	128	21.13	136	26.12-18	140		
19.28	128	21.15-17	136	26.13	69		
19.30-31	128	21.18-21	136	26.16	69		
19.34	128	21.20	43	26.16-18	33		
20.2-6	131	21.22-24	136	26.19-27	139		
20.2-12	131	21.25	136	26.28-29	139		
20.5-6	93	21.27-29	136	26.30-32	11, 139		
20.5-16	104	21.31-38	136	27	94, 139		
20.6		21.39-40	136	27.1	92, 145		
20.6-12	133	22	137	27.1-8	143		
20.7	131, 167	22.1-21	38, 136, 168	27.1–28.14	93		
20.7-12	131			27.1–28.16	104		
20.5-16	104	22.3-5	136	27.1–28.31	11, 12, 143, 145, 147, 149, 151, 153, 155, 157, 159, 168		
20.9	131	22.4	69				
20.9-12	167	22.6	136				
20.10	131	22.7	69, 136, 137				

27.2-5	145	27.33-36	155	28.16	92, 156
27.2–28.13	23	27.33-38	147, 149,	28.17-22	156
27.6	145		159	28.23	156
27.7-8	145	27.34	148	28.23-28	156
27.9	94, 145,	27.36	149	28.23-31	5, 10
	159	27.38	150	28.24	156
27.9-20	145	27.39	151	28.25-28	38, 44
27.10	145, 159	27.39-44	146, 151	28.26-27	156
27.11	146	27.40	151	28.28	98, 157
27.12	146	27.41	151	28.30	92, 157
27.13	146	27.43	151	28.30-31	157, 168,
27.13-44	96	27.43-44	152		169
27.14	146	27.44	152, 159	28.31	11, 45, 153,
27.15-17	146	28.1	152		160
27.18-20	146	28.1-31	152		
27.18-38	146	28.2	152	**1 Corinthians**	
27.20	152	28.3	152	16.2	131
27.21	147	28.4	153		
27.21-22	146	28.5-10	153	**2 Corinthians**	
27.21-25	146	28.8	148, 153	11.26	70
27.21-26	147	28.9	154	11.27	70
27.21-38	146, 147	28.10-11	154		
27.22-32	147	28.11	154	**Galatians**	
27.26	146	28.11-13	154	2.11-14	76
27.27-29	146	28.13	114		
27.31	147, 152	28.14	155	**Revelation**	
27.32	147	28.15	155, 156	1.10	131

www.ingramcontent.com/pod-product-compliance
Lightning Source LLC
Chambersburg PA
CBHW052044300426
44117CB00012B/1974